Micro- and Nanofluidics for Bionanoparticle Analysis

Micro- and Nanofluidics for Bionanoparticle Analysis

Special Issue Editors

Xuanhong Cheng
Yong Zeng

MDPI • Basel • Beijing • Wuhan • Barcelona • Belgrade

Special Issue Editors

Xuanhong Cheng
Lehigh University
USA

Yong Zeng
University of Kansas
USA

Editorial Office
MDPI
St. Alban-Anlage 66
4052 Basel, Switzerland

This is a reprint of articles from the Special Issue published online in the open access journal *Micromachines* (ISSN 2072-666X) from 2018 to 2019 (available at: https://www.mdpi.com/journal/micromachines/special_issues/Micro_Nanofluidics_Bionanoparticle_Analysis)

For citation purposes, cite each article independently as indicated on the article page online and as indicated below:

LastName, A.A.; LastName, B.B.; LastName, C.C. Article Title. *Journal Name* **Year**, *Article Number*, Page Range.

ISBN 978-3-03921-594-2 (Pbk)
ISBN 978-3-03921-595-9 (PDF)

Contents

About the Special Issue Editors

Xuanhong Cheng is an associate professor in the Department of Materials Science and Engineering and the Department of Bioengineering at Lehigh University. She received her Ph.D. in Bioengineering from University of Washington in 2004, after which she had a postdoctoral training in the Department of Surgery at Massachusetts General Hospital. The focus of her research is to develop microfluidic devices for cell and pathogen analysis. In combination with nanomaterials and biosensors, these microfluidic devices are expected to perform automated sample processing and sensitive analyte detection towards point-of-care uses. She has published more than 60 referred technical papers, has been granted several U.S. patents and has obtained research funding from NIH, NSF, and DoD.

Yong Zeng currently is a Docking Scholar Associate Professor in the Department of Chemistry and KU Cancer Center at the University of Kansas. He received his Ph.D. in Chemistry from the University of Alberta, Canada, completed his postdoctoral training at the University of California, Berkeley, and joined KU as assistant professor in 2012. Working at the interface of chemistry and biology, a primary aim of his research is to develop new enabling tools for liquid biopsy-based cancer diagnosis and precision medicine. His work has led to many highly recognized publications on high-profile journals, including *Nature Biomed. Eng.*, *Chem. Sci.*, and *Angew. Chem. Int. Ed.* He received the J.R. & Inez Jay Award in 2014, and was named as the prestigious Docking Family Faculty Scholar in 2017. He has received several million research funding from National Institute of Health (NIH) and other funding agencies.

 *micromachines* 

Editorial

Editorial for the Special Issue on "Micro- and Nanofluidics for Bionanoparticle Analysis"

Xuanhong Cheng [1,*] **and Yong Zeng** [2,*]

[1] Materials Science and Engineering/Bioengineering, Lehigh University, 5 E. Packer Ave, Bethlehem, PA 18015, USA

[2] Department of Chemistry, University of Kansas, 1251 Wescoe Hall Drive, Lawrence, KS 66045, USA

* Correspondence: xuc207@lehigh.edu (X.C.); yongz@ku.edu (Y.Z.)

Received: 10 September 2019; Accepted: 11 September 2019; Published: 12 September 2019

Bionanoparticles such as microorganisms and exosomes are recognized as important targets for clinical diagnostic and therapeutic applications as well as for food safety and environmental monitoring. Other nanoscale biological particles including liposomes, micelles, and functionalized polymeric particles are widely used in nanomedicine. The recent development of microfluidic and nanofluidic technologies has enabled lab-on-a-chip platforms for separating and analyzing bionanoparticles; however, many challenges exist before these platforms are widely adopted. For example, the complex composition of biological fluids creates a high background for detection. The small dimension and often low concentration of target species require significant amplification of the sensing signal. This special issue collects some recent discoveries and developments of micro- and nanofluidic strategies for the processing and analysis of biological nanoparticles.

Eight papers are published in this special issue and cover the design of microfluidics for bioseparation [1–6], incorporation of sensors with microfluidics for particle detection [6,7], and modular devices with both functions [1,6,7]. Nanomaterials have been combined with microfluidics for both separation [5] and sensing [7] purposes. The special issue also includes a review on programmable paper-based microfluidics [8].

The collection of papers describe a wide spectrum of separation and sensing mechanisms. For example, Wang et al. [5] incorporated nanoporous membranes into microfluidic channels, and demonstrated the separation and enrichment of virus from biological fluids through nanofiltration. Yang et al. [6] took advantage of particle inertia in curved microfluidic channels to separate and contain particles in micro-tanks, which were further resolved by optical diffraction. Cheng et al. [7] coated a surface acoustic wave sensor with oxidized mesoporous carbon nanospheres to trap chemical compounds in smoke, where the coating increased the sensitivity and selectivity of the acoustic wave sensor. Boltz et al. [1] designed a split-flow device to perform continuous, in-line quality control of nanoparticle synthesis. Nanoparticles were detected by incorporating a fluorescence detector with the side channel. Soum et al. [4] fabricated conductive electrodes and dielectric films on a paper-based microfluidic chip to facilitate droplet manipulation based on electrowetting on dielectric materials. Luo et al. [3] studied the separation of non-magnetic particles in ferrofluids under a magnetic field. Liao et al. [2] employed optically induced dielectrophoresis to separate cancer cells from blood. The review paper by Soum [8] examined various paper-based microfluidics that are programmable and discussed the utility of such devices for the detection of biomarkers.

We appreciate all of the authors who submitted papers to this special issue. We would also like to thank all of the reviewers for dedicating their time to help improve the quality of the submitted papers. The collection of papers will hopefully bring out more innovative ideas and fundamental insights to overcome the hurdles faced in the separation and analysis of bionanoparticles.

Conflicts of Interest: The author declares no conflict of interest.

References

1. Bolze, H.; Erfle, P.; Riewe, J.; Bunjes, H.; Dietzel, A.; Burg, T.P. A Microfluidic Split-Flow Technology for Product Characterization in Continuous Low-Volume Nanoparticle Synthesis. *Micromachines* **2019**, *10*, 179. [CrossRef] [PubMed]
2. Liao, C.-J.; Hsieh, C.-H.; Chiu, T.-K.; Zhu, Y.-X.; Wang, H.-M.; Hung, F.-C.; Chou, W.-P.; Wu, M.-H. An Optically Induced Dielectrophoresis (ODEP)-Based Microfluidic System for the Isolation of High-Purity $CD45^{neg}$/$EpCAM^{neg}$ Cells from the Blood Samples of Cancer Patients—Demonstration and Initial Exploration of the Clinical Significance of These Cells. *Micromachines* **2018**, *9*, 563. [CrossRef] [PubMed]
3. Luo, L.; He, Y. Magnetically Induced Flow Focusing of Non-Magnetic Microparticles in Ferrofluids under Inclined Magnetic Fields. *Micromachines* **2019**, *10*, 56. [CrossRef] [PubMed]
4. Soum, V.; Kim, Y.; Park, S.; Chuong, M.; Ryu, S.R.; Lee, S.H.; Tanev, G.; Madsen, J.; Kwon, O.-S.; Shin, K. Affordable Fabrication of Conductive Electrodes and Dielectric Films for a Paper-Based Digital Microfluidic Chip. *Micromachines* **2019**, *10*, 109. [CrossRef] [PubMed]
5. Wang, Y.; Keller, K.; Cheng, X. Tangential Flow Microfiltration for Viral Separation and Concentration. *Micromachines* **2019**, *10*, 320. [CrossRef]
6. Yang, N.; Chen, C.; Li, T.; Li, Z.; Zou, L.; Zhang, R.; Mao, H. Portable Rice Disease Spores Capture and Detection Method Using Diffraction Fingerprints on Microfluidic Chip. *Micromachines* **2019**, *10*, 289. [CrossRef]
7. Cheng, C.-Y.; Huang, S.-S.; Yang, C.-M.; Tang, K.-T.; Yao, D.-J. Detection of Cigarette Smoke Using a Surface-Acoustic-Wave Gas Sensor with Non-Polymer-Based Oxidized Hollow Mesoporous Carbon Nanospheres. *Micromachines* **2019**, *10*, 276. [CrossRef]
8. Soum, V.; Park, S.; Brilian, A.I.; Kwon, O.-S.; Shin, K. Programmable Paper-Based Microfluidic Devices for Biomarker Detections. *Micromachines* **2019**, *10*, 516. [CrossRef] [PubMed]

Review

Programmable Paper-Based Microfluidic Devices for Biomarker Detections

Veasna Soum, Sooyong Park, Albertus Ivan Brilian, Oh-Sun Kwon and Kwanwoo Shin *

Department of Chemistry, Institute of Biological Interfaces, Sogang University, Seoul 04107, Korea
* Correspondence: kwshin@sogang.ac.kr

Received: 1 July 2019; Accepted: 1 August 2019; Published: 2 August 2019

Abstract: Recent advanced paper-based microfluidic devices provide an alternative technology for the detection of biomarkers by using affordable and portable devices for point-of-care testing (POCT). Programmable paper-based microfluidic devices enable a wide range of biomarker detection with high sensitivity and automation for single- and multi-step assays because they provide better control for manipulating fluid samples. In this review, we examine the advances in programmable microfluidics, i.e., paper-based continuous-flow microfluidic (p-CMF) devices and paper-based digital microfluidic (p-DMF) devices, for biomarker detection. First, we discuss the methods used to fabricate these two types of paper-based microfluidic devices and the strategies for programming fluid delivery and for droplet manipulation. Next, we discuss the use of these programmable paper-based devices for the single- and multi-step detection of biomarkers. Finally, we present the current limitations of paper-based microfluidics for biomarker detection and the outlook for their development.

Keywords: paper-based microfluidic device; flow control; droplet actuation; multi-step assay; biomarker detection; digital microfluidic device

---

1. Introduction

Because of cost, disposable, paper-based microfluidic devices have become an attractive tool for point-of-care testing (POCT) and medical screening in the developing world. Paper-based, continuous-flow microfluidic (p-CMF) devices were first invented in 1949 by Müller and Clegg [1]; however, no proof of concept for the use of such devices for portable, onsite detection in biosensing applications was reported until 2007 when Whitesides' group published their report [2]. Not only are p-CMF devices cost-effective and eco-friendly, they can also transport fluid samples via capillary action. Because capillary action in a paper channel is passive, it limits the flexibility of fluid handling. Being the focus of research by many academics, p-CMF devices were soon developed so that they could be programmed to transport fluid samples to target locations within a set time. These programmable p-CMF devices can automatically transport fluid samples in sequence for the whole of detection protocols. Therefore, they can be used in a wide range of analytical assay applications involving both single- and multi-step detection protocols.

Another type of paper-based microfluidic device is the paper-based digital microfluidic (p-DMF) device, which was introduced by Shin's group [3] and Wheeler's group [4] in 2014, even though the first DMF had been reported by Pollack et al. in 2000 [5]. P-DMF devices transport a fluid sample (droplet) by using the electrowetting on dielectric (EWOD) technique in which a droplet is actuated by applying an electric field to electrode arrays coated with a hydrophobic dielectric layer [6,7]. Although p-DMF devices are based on the same principle as other DMF devices, they have advantageous features such as portability and the capability to deliver, delay, merge, and split small-volume samples. Because p-DMF devices are inexpensive and easy to fabricate, they are more suitable for POCT than DMF devices fabricated on plastic films [8–10], printed circuit boards (PCBs) [11,12], and glass substrates [13,14] are.

Since the initial reports on p-CMF and p-DMF devices, they have become attractive research topics and have been developed for real-world applications in biomarker detection.

Easy, low-cost fabrication is a major advantage of paper-based microfluidic devices [15]. The p-CMF is commonly fabricated by patterning a hydrophobic material (fluid barrier) on the surface of hydrophilic paper [1,2,16–18] and cutting out a channel from the hydrophilic paper [19–22]. The paper channel of the p-CMF device is inherently a porous matrix that provides continuous capillary action for fluid transport. The capillary action on the paper surface can be controlled by way of physical and chemical modifications [23–25]. Through control of the capillary action, the p-CMF device can be programmed for challenging detection protocols to manage the fluid flow better [25,26]. On the other hand, p-DMF devices are fabricated by patterning conductive electrode arrays on paper substrates and then coating the electrode arrays with a hydrophobic-dielectric layer [3,4]. Because the operation of a p-DMF device is based on the EWOD technique, an electrical switching system and a control software interface are needed to program the fluid transport [27,28].

As analytical platforms, paper-based microfluidic devices are commonly used to transport fluid samples and to store chemical reagents for the colorimetric and electrochemical detection of target analytes [29–32]. The traditional p-CMF is limited by low sensitivity and by difficulty in achieving multi-step assays with automatic processes [25]. Furthermore, p-DMF devices still require a portable electrical switching system and a controlled software interface for the assay protocols [33]. Programming the fluid transport (delay, mix, merge, split, valve, and sequential fluid delivery) in the p-DMF device provides flexibility of fluidic manipulation and opens a wide range of analytical assays for a better detection result.

In this review, we describe the recent advances in programmable paper-based microfluidic devices from their fabrication methods to their applications for biomarker detection, as schematically shown in Figure 1. For p-CMF and p-DMF devices, we present the methods used for their fabrication and the techniques used to control fluid transport in a programmable way. We give some examples of lab-on-a-chip applications using programmable paper-based microfluidic devices. We end this review with a discussion of the current limitations on and the future aspects for the use of programmable paper-based microfluidic devices in real-world applications.

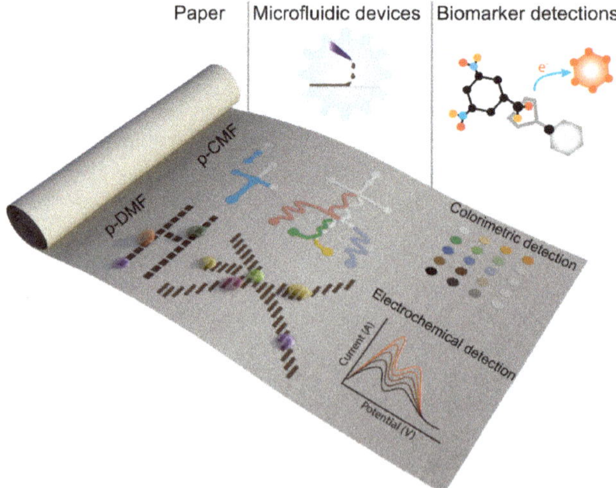

Figure 1. Schematic illustration of a programmable microfluidic device (paper-based continuous-flow microfluidic (p-CMF) and paper-based digital microfluidic (p-DMF)) fabricated on a sheet of paper. Various developed methods can be used to program these devices to manage a fluid sample for high-sensitivity and multi-step assay detection of biomarkers at point-of-care (POC).

2. Fabrication of p-CMF Devices

The fibrous cellulose network in a filter, chromatography paper, and nitrocellulose paper allow fluid to flow by capillary action in a paper matrix. The fluid flow in the paper is mainly influenced by the cohesive and the adhesive forces that are dominated by the surface tension of the fluid and the surface chemistry of the cellulose fiber network, respectively. A paper-based guided channel for fluid flow can be fabricated by turning a sheet of paper into a paper-based microfluidic platform that can be used to manipulate the fluid sample. The common methods used for the fabrication of paper-based microfluidic devices, such as patterning a hydrophobic barrier, cutting out a channel, layer-by-layer assembly, and laser treatment, can be found in various review papers [15,23,34]. These methods allow paper-based microfluidic devices to be fabricated at low-cost and in a simple way.

Patterning a hydrophobic fluid barrier in a fibrous cellulose network, which was introduced in 1949 by Müller and Clegg [1], was the first method used to fabricate paper-based microfluidics. Creating a hydrophobic fluid barrier involves two principles: physically blocking the porous networks in the paper, and chemically modifying the surfaces of the cellulose fibers in the paper. To block the pores in the paper physically, one has to deposit a hydrophobic material into the matrix of the paper to create fluidic barriers (Figure 2) [15,23]. Paraffin film is a pre-made, low-cost material that is widely used in laboratories to shield containers so as to avoid leakage. Because paraffin is hydrophobic, it can be used to create the hydrophobic barriers needed for the fabrication of p-CMF devices. Depending on the design of the microfluidic channel, one cuts paraffin out from its film and hot-laminates it so that it melts on the surface of the filter paper. The melted paraffin penetrates into the porous network in the paper to create hydrophobic barriers, and the area unoccupied by paraffin remains as a hydrophilic channel; as a result, a p-CMF device can be fabricated [16,35].

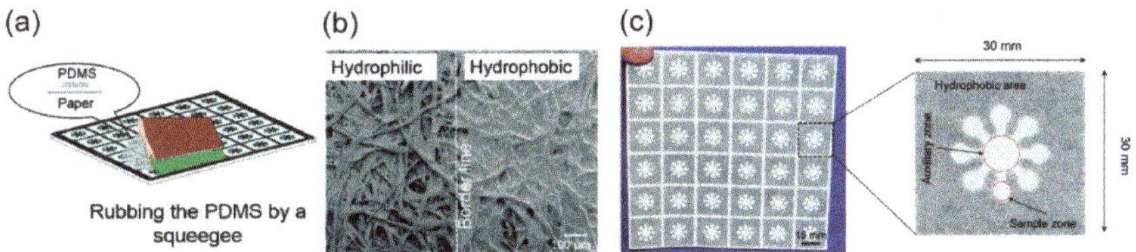

Figure 2. Patterning hydrophobic barriers on paper for fabricating p-CMF devices. (**a**) Screen-printing of polydimethylsiloxane (PDMS ink for patterning PDMS-hydrophobic patterns, (**b**) scanning electron microscope (SEM) images of paper (left) and PDMS-paper (right), and (**c**) fabricated p-CMF devices. Reprinted with permission from the authors of [36]. Copyright 2015 The Royal Society of Chemistry.

Hydrophobic wax is a common material used in the fabrication of p-CMF devices because it is cheap and printable. It can be printed by using a wax printer to generate wax patterns on the surface of paper [17,37–39]. After patterning, the printed wax has to be heated at about 100 °C to melt so that it can penetrate the paper matrix to block the porous network in the paper. For the fabrication of p-CMF devices, polydimethylsiloxane (PDMS) and polymethylmethacrylate (PMMA) can be used to block the porous networks in paper [36,40]. An aqueous solution of PDMS and PMMA can flow into the matrix of paper immediately after screen-printing; however, the printed patterns require a post-treatment to remove solvents to obtain hydrophobicity. PDMS patterns have to be cured by drying at about 120 °C while PMMA patterns can be left to dry in the ambient. Titanium oxides (TiO_2) have been used as hydrophobic agents to block the pores in paper [41]. They were formulated as aqueous ink and deposited into the paper by direct handwriting with a correction pen. Another method, chemical surface modification, is performed by introducing a new material to bind to the cellulose paper chemically, thus forming hydrophobic surfaces. The common materials used in the modifications are alkyl ketene dimer (AKD) [42,43] and commercial permanent marker ink [18,44,45]. In the fabrication of p-CMF devices, Inkjet printing and pen-plotting are used for patterning AKD and the permanent marker ink, respectively, onto the surface of the paper [42–44]. Trichlorosilane (TCS) is deposited on paper by using chemical vapor deposition (CVD) to create fluidic barriers in p-CMF devices [46].

Cutting a piece from porous paper, such as a filter or chromatography paper, is a straightforward method for fabricating paper-based microfluidic devices. It does not require any external material to be deposited in the paper matrix as the above methods do because the porous networks in the paper end at the cutting edge (Figure 3b). The part of the paper cut from the paper sheet is directly used as the microfluidic material (Figure 3c). This method mainly relies on cutting with a blade [21,22] or a laser, for example [20,47]. Cutting the paper with a blade is called the contact-cutting mode and can be performed manually by using scissors or a blade with a mask. The cutting can also be done digitally by using a digital cutting plotter with its blade. The cutting resolution depends on the sharpness of the blade and the toughness of the paper. On the other hand, cutting the paper with a laser is called the non-contact cutting mode and provides fast, high quality cutting with programmable control (Figure 3a).

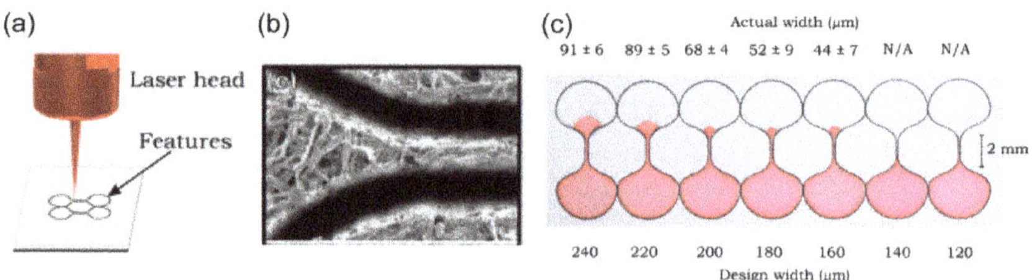

Figure 3. Cutting and shaping paper in the fabrication of p-CMF devices. (**a**) The use of a laser beam to cut and shape paper to form a large paper channel, (**b**) SEM image of the paper channel formed by using the laser, and (**c**) prepared p-CMF devices with various channel widths. Reproduced with permission from the authors of [20]. Copyright 2018 under the terms and conditions of the Creative Commons Attribution.

Paper-based microfluidic devices can be fabricated by assembling the cut paper, the paper or the adhesive film, the paper to form a three-dimensional (3D) channel (Figure 4). Stacking, origami, and lamination can be used to assemble each layer of the microfluidic device [48–53]. This fabrication method relies on the abilities to pattern a hydrophobic fluid barrier and to use a cutting method to generate each layer before the layers are assembled together. The assembling of the layers of the fabricated microfluidic device in the stacking and the origami methods requires a holder to ensure that the layers stay in contact with one another. The microfluidic devices fabricated by using the lamination technique do not require a holder because each layer can bind together due to the adhesive material coated on the laminating films. Molding is another way to fabricate a paper channel. In this method, a mold is created and filled with a mixture of cellulose powder and water [54]. A paper channel with cellulose networks based on the shape of the mold is generated after drying.

Embossing is a relatively new and low-cost fabrication method for p-CMF. Because paper is soft and porous, it can be compressed to form any desired geometry. Using a compressor to apply sufficient pressure to the paper with a stamp, researchers can rapidly form paper channels (Figure 5a–c). The paper channel can either be a compressed or molded area [48,55,56]. However, the pristine channel requires hydrophobic barriers to work as guided barriers (Figure 5d). The hydrophobic barriers can be created by silanization on the compressed channels [48] and applying wax to the boundaries of the molded channels [35,55,56]. During the embossing process of a hot-embossing method, hydrophobic barriers and paper channels can be made, so no post-treatment is required for the fabrication [56].

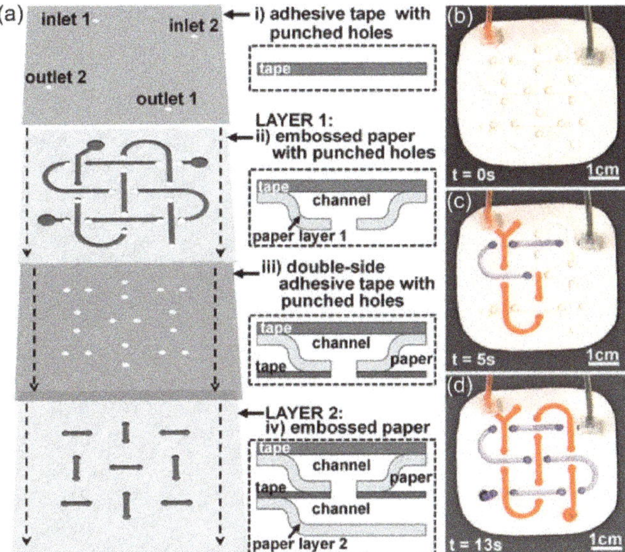

Figure 4. (**a**) Fabrication of a 3D p-CMF device by assembling components such as adhesive tape and embossed paper. (**b**) The fabricated 3D p-CMF device before fluidic loading and (**c**) after 5 s of fluidic loading. (**d**) The flow of the fluid at the ends of the channels. Reprinted with permission from the authors of [48]. Copyright 2014 American Chemical Society.

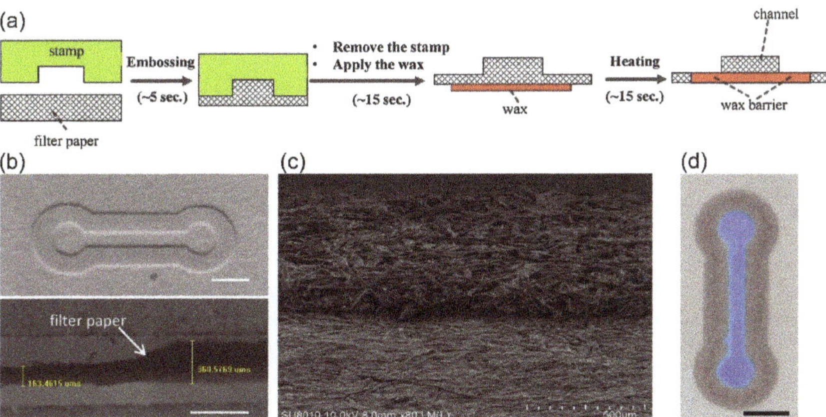

Figure 5. (**a**) Schematic of the fabrication process, including embossing, for the p-CMF device. (**b**) Images of the top (top) and cross-sectional (bottom) views of the embossed paper channel. (**c**) SEM image of the boundary of the embossed paper channel before applying wax. (**d**) p-CMF device after applying the wax and loading it with blue ink. Scale bars represent 5 mm. Reprinted with permission from the authors of [55]. Copyright 2018 Elsevier B.V.

A laser beam can cause a hydrophobic surface to become a hydrophilic surface. Based on the hydrophilic-hydrophobic contrast principle in the fabrication of p-CFM devices, one can use laser light to irradiate hydrophobic paper to produce hydrophilic channels in the surface of that paper (Figure 6). First, pristine paper has to be coated with a hydrophobic agent, for instance, octadecyltrichlorosilane (OTS) n-hexane [57], TiO_2 [58,59], a photopolymer [60,61], or a hydrophobic so-gel [62,63], to generate hydrophobic paper. Once the hydrophobic paper has been made, a designed area on the hydrophobic surface can be exposed to laser irradiation to create hydrophilic channels. The hydrophilic channels allow fluid to flow while the remaining hydrophobic area functions as a fluid barrier. Summary of some published fabrication methods of p-CMF devices is shown in Table 1.

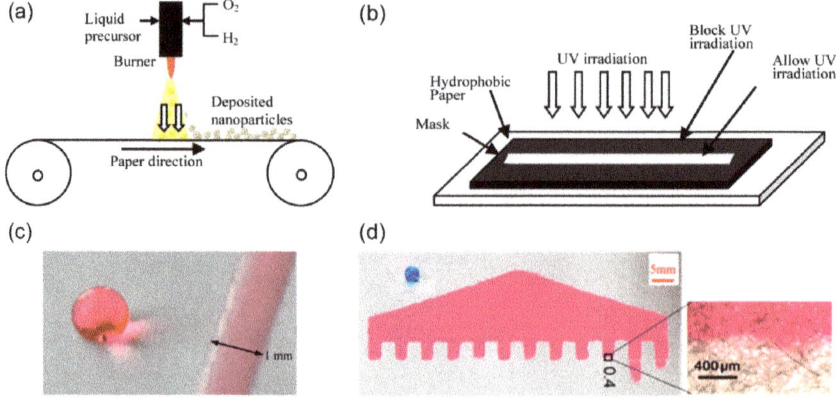

Figure 6. (**a**,**b**) Schematic showing the use of a laser to irradiate a hydrophobic surface to fabricate a p-CMF device. (**c**) Water droplet on a hydrophobic surface and the spreading of water on a hydrophilic surface that had been subjected to UV irradiation. Reprinted with permission from the authors of [58]. Copyright 2014 American Chemical Society. (**d**) A typical design of a fabricated p-CMF device, and a high magnification image of the boundary between the hydrophobic and the hydrophilic surfaces. Reprinted with permission from the authors of [62]. Copyright 2018 American Chemical Society.

Table 1. Summary of published fabrication methods, materials, advantages, and disadvantages of for p-CMF devices.

Method [References]	Materials	Advantages	Disadvantages
Hot laminating [16]	Paraffin film	Simple and low-cost	Requires a cutter and hot-laminator
Wax printing [17,37–39]	Wax	Rapid and low-cost; no mask needed; programmable printing; mass production	Requires cure at high temperature
Screen printing [36,40]	Polydimethylsiloxane (PDMS); polymethylmethacrylate (PMMA)	Rapid and low-cost; can print with high viscosity ink	Requires cure at high temperature and screen mask
Photolithography [2,57]	Photoresist; octadecyltrichlorosilane (OTS)	High resolution; mass production	Requires clean room facilities
Direct handwriting [41]	Titanium oxide (TiO$_2$)	Rapid and low-cost	Heavily depends on writers' skill; difficult to mass produce
Inkjet printing [42–44]	Alkyl ketene dimer (AKD); commercial permanent marker ink	Customizable design; programmable printing	Requires low viscosity ink; nozzle clogging
Plotting [18,45,64]	PDMS; commercial permanent marker ink	Simple operation; low-cost	Requires a customized dispenser
Chemical vapor deposition (CVD) [46]	Trichlorosilane (TCS)	Rapid and simple process	Requires vacuum pump and mask

Table 1. *Cont.*

Method [References]	Materials	Advantages	Disadvantages
Cutting [20–22,47]	–	No hydrophobic agent is required; high resolution (laser cutter)	The resolution depends on the sharpness of a cutting blade; Expensive (laser cutter)
Layer-by-layer assembling [48–53]	$R^F SiCl_3$, wax, photoresist; tape	Can create 3D channel; multiple layer channels	Requires a cutter, tape, and holder
Molding [54]	Cellulose powder	No hydrophobic agents are required	Requires molding
Embossing [35,48,55,56]	$R^F SiCl_3$; wax; Paraffin film,	Can create 3D channel; simple and low-cost	Requires high pressure and molding
Laser treatment [57–63]	Octadecyltrichlorosilane (OTS); TiO_2; photopolymer; hydrophobic so-gel	Can control surface energy of paper channel	Requires large area of hydrophobic coating on paper

3. Fabrication of p-DMF Devices

Digital microfluidic devices actuate a fluid droplet based on the EWOD technique, which requires a set of conductive electrode arrays, a dielectric layer, and a hydrophobic layer. The conductive electrode array allows the applied voltage to induce an electrostatic force at the interfaces between the dielectric layer and conductive liquid while the hydrophobic layer minimizes the surface energy and the surface friction. The characteristics of the conductive electrode arrays and the hydrophobic-dielectric layer are important parameters that influence the performance and the robustness of a DMF device. We should note that the fabrication cost of those components must be low if the devices are to be used for POCT. Because DMF devices can be fabricated on paper substrates, the so-called p-DMF devices, they should be cheap, disposable, and suitable for low-resource laboratories.

The fabrication of a p-DMF device starts from the patterning of conductive electrode arrays on a paper substrate. The patterning of conductive electrode arrays can be done by using photolithography or an etching method, which requires clean-room facilities. For simple, low-cost patterning of conductive electrode arrays for p-DMF devices, Shin's group introduced the Inkjet printing method [3,28]. They printed carbon nanotube (CNT) ink by using an office Inkjet printer to pattern conductive CNT electrodes on photo paper (Figure 7a). Fobel et al. also used the Inkjet printing method for patterning conductive electrode arrays on photo paper [4]. Their conductive electrode arrays were printed using silver nanoparticle (AgNP) ink, so they could obtain electrodes with higher conductivity compared to the ones printed from CNT ink. The Inkjet-printed electrodes are thin (about submicron to a few microns thick), which is suitable for thin dielectric layer coating, the next step in the fabrication of p-DMF devices. Another printing method, screen printing, can be used for patterning conductive electrodes on paper [65,66]. By using this method with silver-based and carbon-based ink, Yafia et al. printed multiple conductive electrode arrays on paper [65]. Screen-printed electrodes are generally thick ($\geq$10 μm) because the ink used for printing has high viscosity. With high-pressure air, Abadian and Jafarabadi-Ashtiani generated conductive electrode arrays on paper by spraying graphite onto paper that had been covered with a mask [67]. Jafry et al. introduced a micro-syringe pump to dispense AgNP ink for patterning conductive electrode arrays on filter paper [68]. A contact printing method using a ballpoint pen with a digital plotter was recently proposed and used for patterning AgNP electrode arrays on photo paper (Figure 7d,e) [69]. That method provides an affordable way for printing in a programmable manner. When electrodes are printed in this way, their thicknesses are roughly submicron, and they have high electrical conductivity.

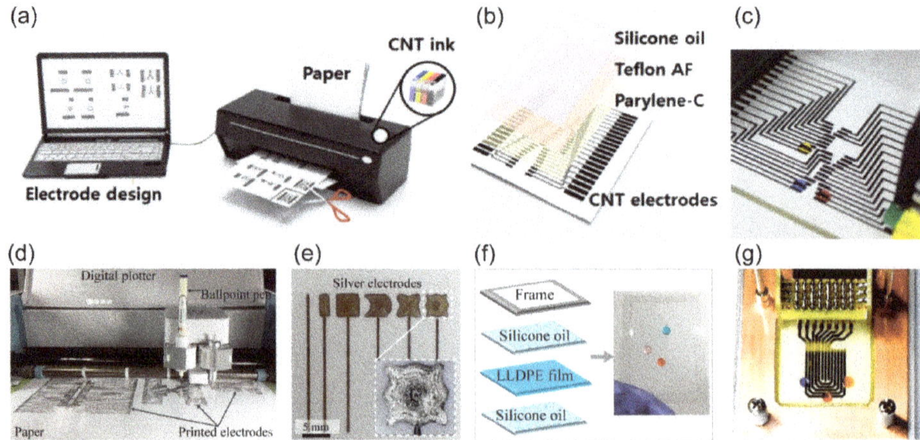

Figure 7. Fabrication of p-DMF devices. (**a**) Patterning of conductive electrode arrays for a p-DMF chip by Inkjet printing of CNT ink on paper. (**b**) Schematic illustration of the components of a p-DMF device: CNT electrodes, Parylene-C dielectric layer, Teflon AF hydrophobic layer, and slippery silicone oil layer. (**c**) A fabricated p-DMF chip. Adapted with permission from the authors of [3]. Copyright 2014 WILEY-VCH Verlag GmbH & Co. KGaA, Weinheim. (**d**) Printing of conductive electrode arrays for a p-DMF chip on paper by using a ballpoint pen with AgNP ink and a digital plotter. (**e**) Printed electrodes, (**f**) a prepared dielectric layer modified with silicone oil, and (**g**) a p-DMF chip. Reproduced with permission from the authors of [69]. Copyright 2019 under the terms and conditions of the Creative Commons Attribution.

After the conductive electrode arrays have been fabricated, a dielectric layer and a hydrophobic or slippery surface are required to make a complete p-DMF device (Figure 7b–g). The material used for the dielectric layer has to have a high breakdown voltage to avoid dielectric malfunction when a high voltage is applied during operation. Parylene-C, a reliable dielectric material for use in electronics, is commonly used as the dielectric layer for p-DMF devices [3,4,28]. It can be deposited on the printed electrode arrays by using chemical vapor deposition (CVD). Another material, a PDMS thin film, can be prepared by spin coating and used as the dielectric layer for a p-DMF chip [68]. For an easy, low-cost method without the need for clean-room facilities, researchers have proposed some affordable materials, such as parafilm [65], adhesive tape [66], and commercial food wrap [69], for use as dielectric layers. Because most of these dialectic materials are hydrophilic or have low hydrophobicity, for better EWOD performance, a hydrophobic and/or slippery layer is required. Teflon is a common hydrophobic material used for hydrophobic coating while silicon oil is used for generating slippery surfaces [4,69,70]. Summary of some published fabrication methods of p-DMF devices is shown in Table 2.

Table 2. Summary of published fabrication methods, materials, advantages, and disadvantages of p-DMF devices.

P-DMF Components	Method [References]	Materials	Advantages	Disadvantages
Conductive electrode arrays	Inkjet printing [3,4,28]	Carbon nanotube (CNT); silver nanoparticle (AgNP)	Rapid; can produce thin electrodes; programmable printing	Nozzle clogging
	Screen printing [65,66]	Carbon; silver	Rapid and low cost; can print high viscosity ink	Requires a mask; produces thick electrodes
	Spraying [67]	Graphite	Simple and low cost	Requires mask; low resolution
	Micro-syringe dispensing [68]	AgNP	Can print high viscosity ink	Complicated setup
	Ballpoint pen printing [69]	AgNP	Simple and low cost; can produce thin electrode	Requires a customized ballpoint pen
Dielectric-hydrophobic layer	Chemical vapor deposition (CVD) [3,4,28]	Parylene-C/Teflon	High quality coating; controllable coating thickness	Requires clean room facilities; expensive
	Spin coating [68]	Polydimethylsiloxane (PDMS)/silicon oil	Simple process	Difficult to coat large area; excessive material waste
	Taping [66]	Adhesive tape/Nevosil	Simple and low cost	Depends on the quality of a commercial tape
	Wrapping [65,69]	Paraffin film; plastic wrap/silicon oil	Simple and low cost	Depends on the quality of a commercial wrap

4. Strategies for Programming the Delivery of a Fluid Sample in p-CMF Devices

Delivery of fluid samples in the p-CMF device is a passive process and results from capillary flow or wicking flow inside the fibrous cellulose network. This passive flow in the p-CMF device can be described by using the Lucas Washburn equation, which was derived to describe capillary flow in a one-dimensional model in cylindrical tubes. In that equation, the flow distance (x) is proportional to the square root of time ($t^{1/2}$):

$$x = \sqrt{\frac{\gamma r \cos(\theta) t}{2\eta}} \tag{1}$$

where γ, θ, and η are the fluid surface tension, the contact angle of the fluid with the channel wall, and the fluid viscosity, respectively; r and t are the pore radius of the paper and time, respectively. Based on the Lucas Washburn equation, the physical structure and the surface chemistry are the critical microfluidic parameters that control the speed of fluid flow in the p-CMF device. With these concepts, researchers have developed a variety of programmable p-CMF devices by using various methods to control these parameters. Some of the excellent methods, such as variation of the channel dimensions (width, and depth), control of the permeability of the paper, use of pattern fluid-flow regulators, and control of the surface chemistry of the paper channels, have been used to manipulate fluid sample delivery in p-CMF devices.

When the channel width is varied, several parameters must be explored to explain the wicking flow of the fluid in open paper channels. In a paper channel that is thick (~0.70 mm) and has a straight boundary, wicking flow shows no changes when the width of the channel is varied (Figure 8a) [71,72]. However, wicking flow decreases in a thinner channel (0.18 mm) when the width of the channel becomes smaller (Figure 8b) [72]. The decrease in flow speed in a thin channel can be explained by the edge effect of the cut channel where the edges hinder the flow. In a flow effect similar to that for a thin channel with cut boundary, a hydrophobic boundary channel enables a faster flow speed in a thicker channel with a larger width (horizontal orientation) (Figure 8c,d) [71,73–75]. When the channel width and height are increased, the edge effect is significantly reduced over the total area of the channel, which allows faster flow. Another method to manipulate the flow in the hydrophobic boundary channel is to carve the channel either longitudinally or perpendicularly, thus accelerating or de-accelerating the flow, respectively [76]. The longitudinally crafted channels create a flow path for fast laminar flow while the perpendicularly crafted channels resist the flow.

Depending on the capillary force, the characteristics of liquid flow in a paper-based channel are influenced by the permeability of the porous paper. Hydrophobic materials can be used to reduce the pore size of the paper so that the permeability is minimized. Wax can be melted so that it can penetrate the matrix of the porous paper to block some pores, or at least minimize their size, because the permeability of the paper depends on the amount of wax deposited in the paper. As the permeability of the paper channel is reduced, the flow speed of the liquid in the channel is decreased [77,78]. Another material, agarose, can be used to control the permeability of porous paper (Figure 9a). Agarose can penetrate the matrix of the porous paper so that the pore size becomes smaller (600 nm to 250 nm) with increasing concentration of agarose (0.5% to 4.0%) [79]. The smaller pore size leads to reduced permeability, so the flow in the paper matrix is slower (Figure 9b).

Dissolvable materials have been used to delay the flow of fluid in p-CMF devices. Dissolvable sugars (trehalose, sucrose) have been used to block temporarily the pores in the paper channel and enhance the viscosity of the fluid [80,81]. During flow, the fluid dissolves the sugar in the paper channel. The dissolution of sugar delays the flow and enhances the viscosity of the fluid. With increasing amount of sugar in the paper channel, the fluid flow can be dramatically delayed (Figure 10a–c). Because of this effect, dissolvable materials have been used to control the volume of the aqueous sample to be delivered by the p-CMF device. Houghtaling et al. formulated dissolvable bridges made of sugar (trehalose) and used them to automate the delivery of fluid samples [82]. Their dissolvable bridge was mounted between the source pad and the main channel. The volume of fluid that passes the dissolvable bridge is defined by the cross-sectional area and the composition of the bridge and by the choice of the feeder material. Jahanshahi-Anbuhi et al. used water-soluble pullulan films as valves in p-CMF devices to shut off the flow after the pullulan film had been dissolved because a sufficient volume of liquid had passed through it (Figure 10d) [83].

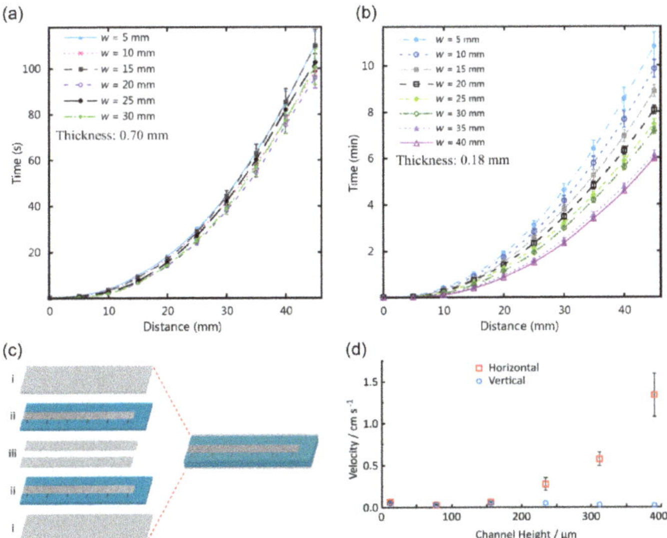

Figure 8. Influence of the geometries of the p-CFM channels on fluidic flow behavior. (**a,b**) Thickness and width of paper channels, (**c,d**) height of channel (the gap between the top and the bottom surface). (**a,b**) Reprinted with permission from the authors of [72]. Copyright 2016 under the terms and conditions of the Creative Commons Attribution. (**c,d**) Reproduced with permission from the authors of [75]. Copyright 2018 The Royal Society of Chemistry.

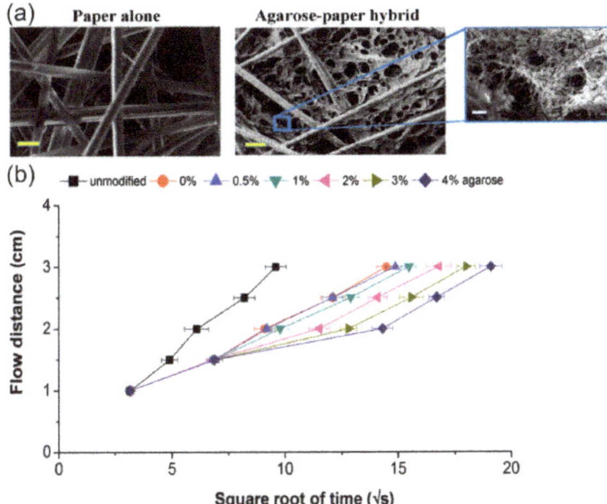

Figure 9. Regulation of the fluid flow in a p-CMF device by controlling the permeability of the paper channel. (**a**) SEM images of paper (left) and of agarose-paper (right). (**b**) Fluid flow distance versus square root of time in paper channels with and without treating with various concentrations of agarose. Reprinted with permission from the authors of [79]. Copyright 2016 WILEY-VCH Verlag GmbH & Co. KGaA, Weinheim.

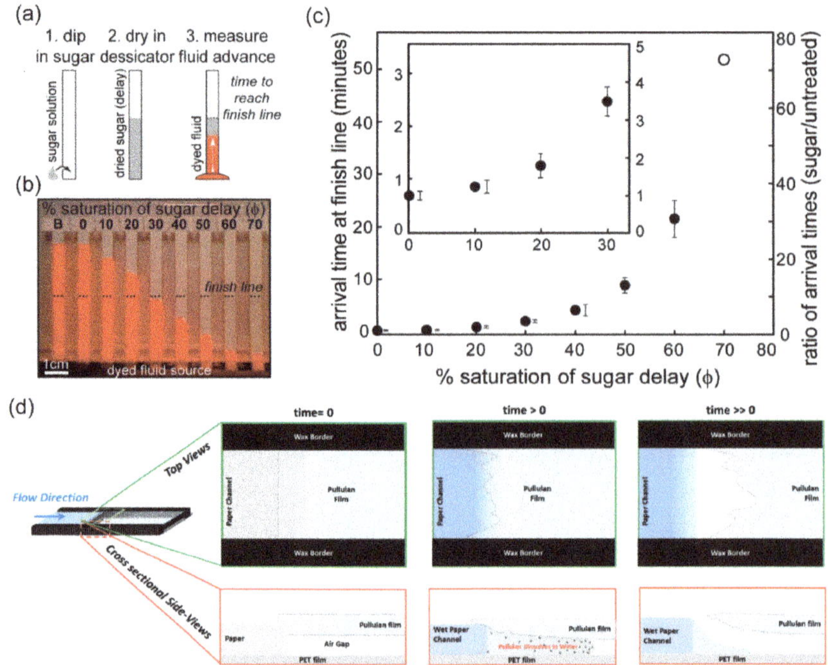

Figure 10. Delay of fluidic flow by incorporating a dissolvable material in a p-CFM device. (**a**) Schematic of a method for preparing a p-CMF device for which time delay is controllable. (**b**) Images of fluidic flow distances at specific times in paper channels modified with various concentrations of sugar. (**c**) Arrival time of fluid at the ends of channels versus the concentration of sugar used in the paper channels. Reproduced with permission from the authors of [81]. Copyright 2013 The Royal Society of Chemistry. (**d**) The pullulan film works as a dissolvable bridge to shut off the flow of liquid when a certain amount of liquid has passed through the bridge. Reproduced with permission from the authors of [83]. Copyright 2014 The Royal Society of Chemistry.

Hydrophobic materials are commonly patterned in the paper channels to regulate the fluid flow in p-CMF devices. The hydrophobic patterns in the paper channel act as obstacles to delay the flow of fluid (Figure 11). Wax pillars were printed on nitrocellulose membrane by Rivas et al. to delay the flow of a fluid sample in the channel [84]. The wax pillars were printed onto the nitrocellulose membrane by using a wax printer; this was followed by heating at 110 °C for 90 min. Using this method, they could slow the flow to 0.18×10^{-3} m/s. Preechakasedkit and co-workers also printed wax as line patterns onto a nitrocellulose membrane channel by using a wax printer [85]. With their design, they could delay the sample delivery time in their channel by approximately 11 s. PDMS patterns were also used as fluidic barriers to slow fluid flow (Figure 11a,b) [86]. The PDMS patterns were created by using a pipette to drop manually 0.1 μL of PDMS solution onto a nitrocellulose membrane. With five patterns, the delivery of the fluid sample could be delayed by 97.2 ± 0.4 s (Figure 11c). He et al. used a laser direct-write (LDW) method to induce a photopolymer to create solid barriers in a paper channel [87]. When the speed of laser direct writing was slower, the polymer patterns created for use as fluidic barriers were thinner. In this way, the fluid flow can be significantly decreased.

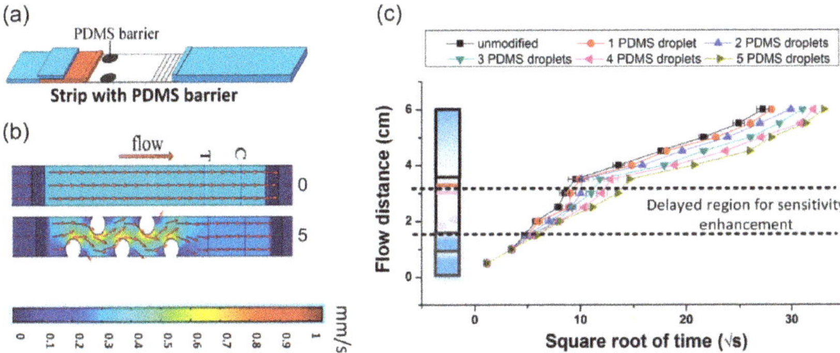

Figure 11. Regulation of the fluid flow in a p-CMF device by using hydrophobic patterns. (**a**) Schematic for the modification of a p-CMF device with PDMS barriers, (**b**) simulation of the flow of fluid in an unmodified and a modified channel, and (**c**) flow distance versus square root of time in the p-CMF devices unmodified and modified with different numbers of PDMS droplets (barrier). Adapted with permission from the authors of [86]. Copyright 2016 American Chemical Society.

Absorbent materials can drag fluid into their matrices through capillary action. The absorbent pads made of porous paper or porous polymer can be used in p-CMF devices to delay the flow of fluid. Toley et al. used paper-based absorbent pads (shunt) mounted on a paper channel to divert liquid flowing in the channel in order to delay delivery of the sample (Figure 12) [88]. By varying the shunt length and thickness, they could achieve 3 to 20 min of time delay (Figure 12b,c). Similar to the above concept, Tang et al. used sponge shunts made of porous polyurethane to decrease the flow rate in a p-CMF device by placing the shunt at a junction between a conjugation pad and a main channel [89]. Akyazi et al. proposed photopolymerized ionogel for delaying the flow of fluid in a p-CMF device [90]. The ionogel solution was deposited on the surface of the paper channel; this was followed by irradiation with UV light for photopolymerization. When the photopolymerized ionogel came into contact with the fluid, it absorbed the fluid into its matrix and swelled. The absorption and the swelling processes delay the time of fluid travel through the paper channel.

Controlling the surface chemistry of the paper channel is an effective method for programming the delivery of a fluid sample by a p-CMF device. When the hydrophobic surface of paper is exposed to laser light, it can become a hydrophilic surface [59]. If the surface wettability of a microfluidic channel can be controlled, both the capillary action in the channel that allows the fluid to flow and the flow speed of the fluid can be controlled (Figure 13). Songok et al. used UV light to irradiate hydrophobic paper (CA ≈ 160°) to create a hydrophilic surface (CA ≈ 15°) for the fabrication of p-CMF devices (Figure 13a) [58,91,92]. They could vary the time delays of fluid flow by varying the geometry of their paper channels (Figure 13b) [91], and they could create fast capillary flow with speeds up to approximately 9 mm/s by adding a hydrophobic top cover on their paper channels (Figure 13c) [92].

Walji and MacDonald studied the influence of temperature on the wicking flow of fluid in a p-CMF device [72]. They found that the fluid flows faster in a paper channel that has been cured at a higher temperature. The 45 °C fluid took about 8 min to flow through a 45-mm channel while the 15 °C fluid took 11 min. Niedl and Beta controlled fluidic transport in p-CMF devices by using heat [93]. Their paper channels were mounted with thermoresponsive hydrogels (NIPAM-AcAm) that stored fluid (Figure 14). At higher temperature, the hydrogels collapse faster and continuously release fluid into the paper channel at a higher rate (Figure 14b,c). They were able to control the release of fluid by varying the temperature in their paper channels.

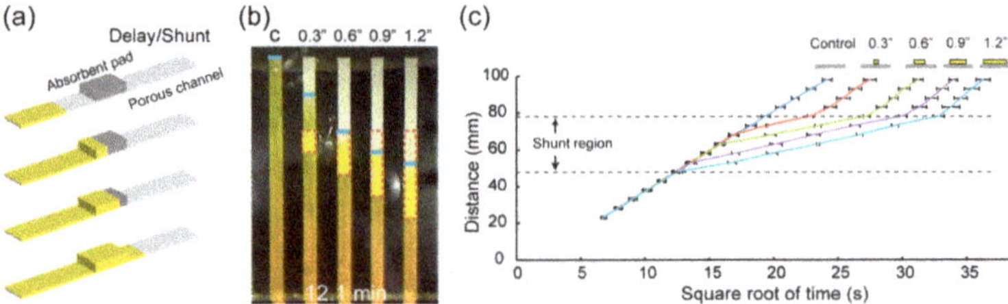

Figure 12. Delay of the fluid flow in a p-CFM device by using absorbent pads. (**a**) Schematics of the flow of fluid in a paper channel modified with an absorbent pad, and (**b**) images of fluid's travel in paper channels modified with absorbent pads of various lengths. (**c**) Flow distance versus square root of the time the fluid is flowing through paper channels modified with absorbent pads of various lengths. Reprinted with permission from the authors of [88]. Copyright 2013 American Chemical Society.

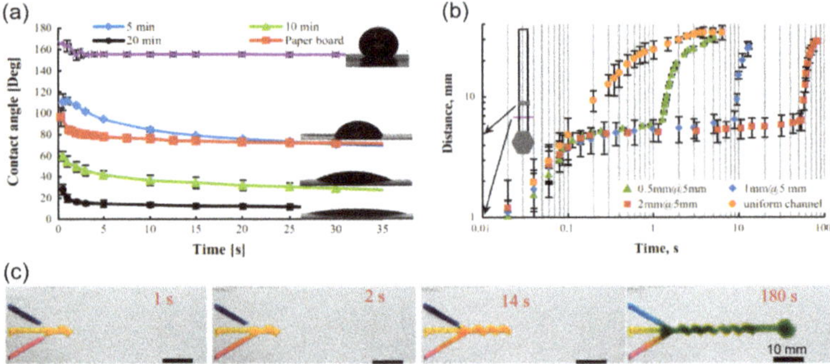

Figure 13. Changing the surface wettability of a paper channel by treating it with UV irradiation to control the flow of fluid in the channel. (**a**) Surface contact angle of a water droplet as a function of time on a paper surface treated with UV irradiation for 0, 5, 10, and 20 min. (**b**) Flow distance of the fluid versus time traveled in paper channels with various geometries due to changes in the wettability of the channel's surface. (**c**) Sequential fluid delivery in the p-MF device. Adapted with permission from the authors of [91]. Copyright 2016 Springer-Verlag Berlin Heidelberg.

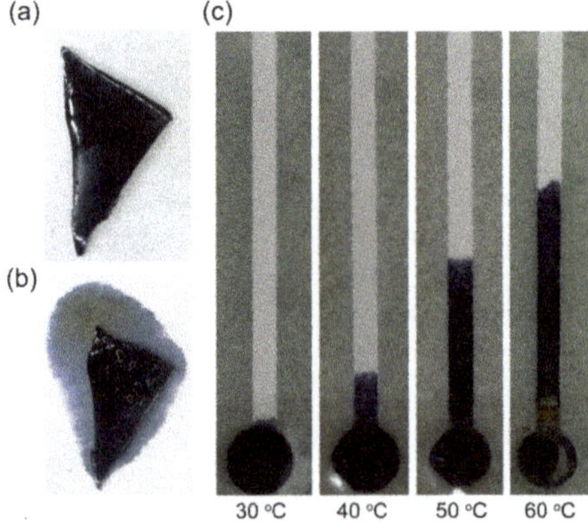

Figure 14. Controlled release of a fluid in a p-CMF device by using a thermoresponsive hydrogel. Fluid (**a**) stored in and (**b**) released from a hydrogel at high temperature. (**c**) Controlling the length of fluid flow in a p-CMF device by heating it at different temperatures. Reproduced with permission from the authors of [93]. Copyright 2015 The Royal Society of Chemistry.

5. Droplet Manipulations in p-DMF Devices

Manipulations, such as transporting, mixing, and merging, of fluid droplets on a p-DMF device are controlled by switching off and on the electrical power applied to the conductive electrode arrays. When a certain voltage is applied, a droplet starts to decrease its contact angle (CA) and reaches a new equilibrium, the so-called wetting equilibrium (Figure 15a). This phenomenon can be explained by the Young–Lippmann equation:

$$\cos \theta_L(V) = \cos \theta_Y(0) + \frac{1}{2\gamma_{lv}} C V^2, \tag{2}$$

where $\theta_L(V)$ and $\theta_Y(0)$ are the CA of a droplet when a voltage V and no voltage are applied, respectively. γ_{lv} is the interfacial tension of conductive liquid (droplet). C and V are the capacitance of the interface and the applied voltage, respectively. When the symmetry of a wetted droplet is broken, it can be moved to a higher wetting surface by an applied voltage (Figure 15b).

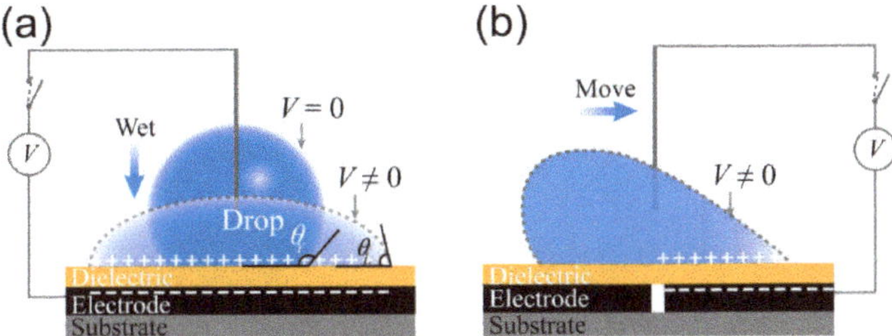

Figure 15. Schematic illustrations of the activation of a fluid droplet on a DMF chip by using the electrowetting on dielectric (EWOD) technique and of droplet wetting when a certain voltage is applied to the system: (**a**) symmetric and (**b**) asymmetric.

Conventionally, when the electrical power applied to operate a p-DMF device is switched on or off, the switch is operated manually or by using the LabVIEW or the MATLAB software [3,68]. When programmable software is used, the handling of the fluid droplets on a p-DMF device is much easier to control than it is when manual switching is used. However, the above systems lack portability because of the number of required devices, such as power supplies, computers, and Arduino circuit boards. Recently, a wireless, portable control system was built to operate a p-DMF device (Figure 16a) [27,28,69]. The control system is based on an Android smartphone operated by the Android OS, which can be installed by using an application called DMF toolbox (version 1.0) (Figure 16b). When the proposed system was used, a p-DMF chip could manipulate fluid droplets in a simple and programmable way (Figure 16c).

Because electrode arrays of p-DMF devices and connection lines are printed on the same side of the paper, they can interfere with each other during a droplet's actuation when they overlap the droplet. Without a feedback system, a droplet can exhibit unexpected movement when an automatic control system is used. To solve this problem, researchers proposed better algorithms for eliminating the interferences during the operation of a p-DMF chip [94,95].

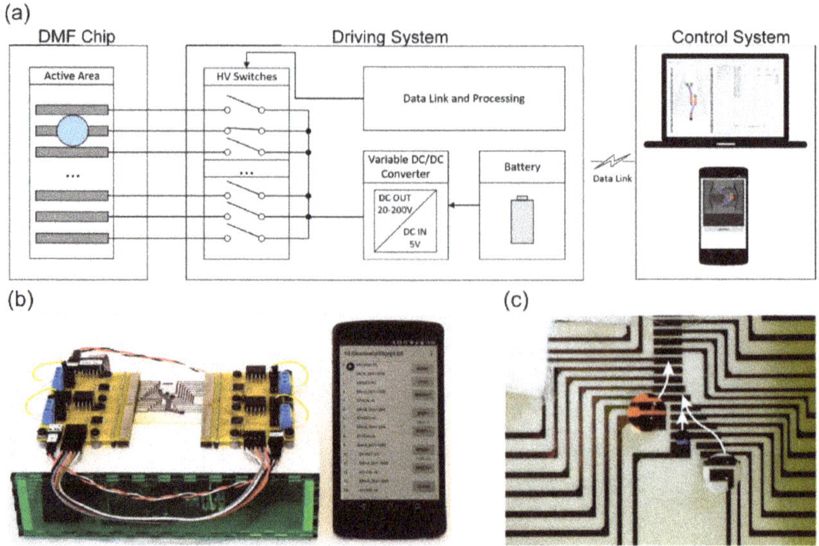

Figure 16. (**a**) Schematic illustration and (**b**) image of the operation system for a p-DMF device. Reproduced with permission from the authors of [27]. Copyright 2017 IEEE. (**c**) A p-DMF chip actuating a droplet by using the operation system. Reproduced with permission from the authors of [28]. Copyright 2017 WILEY-VCH Verlag GmbH & Co. KGaA, Weinheim.

6. Biomarker Detection by Using Programmable p-CMF Devices

That the p-CMF device can transport a sample without the need for an external device makes it very suitable for use as a POCT device. Programming the fluid transport is the ultimate way to achieve an advanced lab-on-a-chip application for the detection of analytes, and it enables single-step and automated multi-step assay protocols with high detection sensitivity. The high detection sensitivity of the p-CMF device is due to the regulation of physical flow and the time delay of the reaction. Because the samples in the p-CMF devices can be programmed for delivery in sequence to the reaction zone, automated multi-step assay detection is achieved.

Colorimetric detection using paper-based microfluidic devices is a common method for detecting target analytes [96,97]. Because the result can be read qualitatively and quantitatively by using the naked eye, this technique has emerged as a prime candidate for use in POCT. However, the detection sensitivity in conventional p-CMF devices is relatively low. Some excellent methods [98] to amplify the detection signal in p-CMF devices, such as the use of metal (Ag, Au, Fe_3O_4/Au) nanoparticles [99–101], catalysts (horseradish peroxidase (HRP), Pt nanocrystals [102,103], and chitosan to modify the paper's surface [104], have been reported. Moreover, regulating the flow behavior of fluid samples by using a controllable p-CMF device is a straightforward method for enhancing the signal in colorimetric detection (Figure 17). When fluidic barriers, such as wax or PDMS patterns, are introduced into the paper matrix in p-CMF devices, the flow behavior of the fluid sample can change from laminar to turbulent flow, which would improve the sensitivity for detection of analytes by up to threefold [84–86]. Absorbent pads (shunts, sponges) have been used for decreasing the flow rate of fluid samples in p-CMF devices to achieve higher signals when detecting nucleic acids [79,86,89]. Designing a p-CMF device to have spatial constrictions of the flow path may lead to a slower flow rate of the fluid sample at the detection zone, thereby allowing more reactants to be bound together; therefore, the signal intensity should be increased [105].

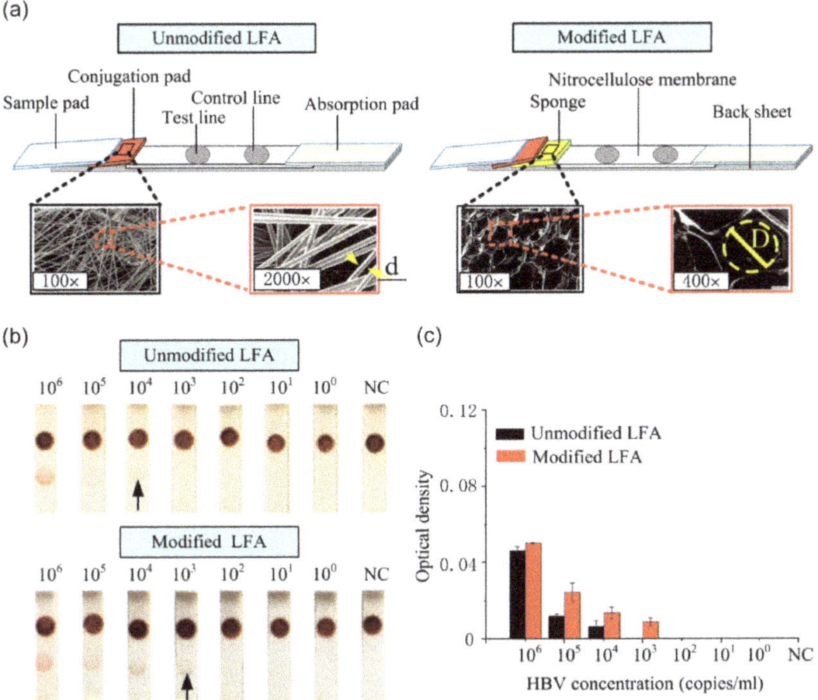

Figure 17. Increasing the detection sensitivity by regulating the flow of the fluid sample in a p-CMF device. (**a**) For lateral flow assay, a sponge was incorporated into a p-CMF device to decrease the flow rate of the fluid sample. (**b,c**) Comparison of the colorimetric detection signals of hepatitis B virus (HBV) for the unmodified and the modified p-CMF devices. Adapted with permission from the authors of [89]. Copyright 2017 under a Creative Commons Attribution.

Most conventional p-CMF devices have the limitation that they cannot automatically transport fluid samples in sequence while most assay protocols for detecting biomarkers require multi-step reactions. To detect those biomarkers using multi-step assay protocols, researchers have developed methods for programming sequential sample delivery in p-CMF devices without requiring valves and pumps. Without the need for these additional devices to control the sample transport, programmable p-CMF devices have become attractive microfluidic platforms for biomarker detection in POCT. Absorbent pads made of cellulose can be placed on the paper channel to divert fluid into it and delay fluid flow in a p-CMF device [88]. If the thickness and the length of the absorbent pads are increased, the fluid flow delay can be increased from 3 min to 20 min. With this control technique, transport of fluid samples in sequence to a detection zone has been realized for the detection of the malaria protein *Pf* HRP2. By varying the amount of sugar (sucrose) dissolved in the channel of a p-CMF device, the time delays of fluid sample delivery could be varied from minutes to nearly an hour (Figure 18) [81]. Because a higher content of sugar in the channel provided a lower fluid-flow speed (longer delay time), the sugar contained in the paper-based channels was varied to allow fluid samples to be delivered to a detection zone in sequence (Figure 18a,b). A multi-step assay for detection of malaria was successfully performed by using a p-CMF device (Figure 18c). A controllable p-CMF device was fabricated by shining laser light onto a filter paper coated with a photo-polymerizable polymer to create a solid barrier. In this p-CMF device, the fluid flow could be delayed from a few minutes to over half an hour by varying the number and the thicknesses of the polymer barriers [87]. This fabricated p-CMF device could be programmed to deliver fluid samples and reagents sequentially for automated multi-step detection of C-reactive protein (CRP). A p-CMF device was fabricated by layering dry pullulan films containing reagents on paper and was successfully used for sequential sample deliveries for multi-step assays for the detection of pH, drugs (methamphetamine-like compounds), and intracellular bacterial enzymes (secondary amines) [106].

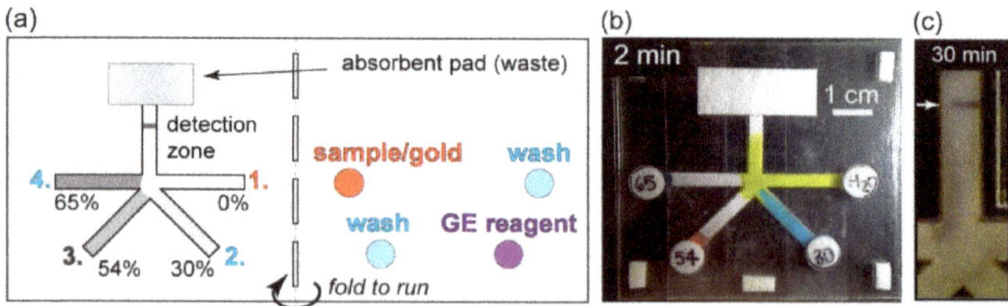

Figure 18. Automated multi-step assay by using a p-CMF device with programmable delivery of the fluid samples. (**a**) Schematic design of a p-CMF for sequential sample delivery in a multi-step assay protocol for detection of malaria. (**b**) Image of sequential sample delivery in the p-CMF device and (**c**) colorimetric detection of malaria, which was achieved by using the programmable p-CMF device. Adapted with permission from the authors of [81]. Copyright 2013 The Royal Society of Chemistry.

7. Biomarker Detection by Using Programmable p-DMF Devices

P-DMF devices, which have performances similar to those of other DMF devices, are novel microfluidic platforms because they can be used to manipulate small-volume droplets independently without the need for channels, pumps, and microvalves; thus, this type of microfluidic device, the p-DMF device, is most suitable for multi-step assay protocols, as well as for other single-step assay protocols [33,107]. In addition to their abilities to transport fluid and delay its flow, p-DMF devices can mix two or more samples effectively to allow homogeneous mixing between a sample and a reagent. Moreover, the p-DMF device is cost effective and disposable, which makes it more suitable for POCT than other DMF devices on glass or PCB. Although the use of p-DMF devices for chemical and biological assays is less popular than the use of other DMF devices fabricated on glass, plastic, and PCB, the use of p-DMF chips as platforms for detecting biomarkers has shown significant progress.

Simple assay protocols for colorimetric detection of glucose by using a p-DMF chip as a microfluidic platform was demonstrated [66]. In the detection, the p-DMF chip was mainly used for mixing the fluid samples and transporting the mixed sample to a p-CMF channel where the detection reagent was immobilized. Another group used a p-DMF chip as a microfluidic platform for sample preparation for protein digestion processes [108]. After protein digestion protocols (disulfide bond reduction, alkylation, buffering, and tryptic digestion) had been conducted using the paper chip, the samples were successfully analyzed by using a MALDI-TOF mass spectrometer to identify the proteins. A programmable, portable p-DMF chip was introduced for multiple electrochemical detections of biomarkers (Figure 19) [28]. Controlled by an Android smartphone via a wireless system, the p-DMF chip was programmed for multi-step assay protocols (Figure 19a). The fluid samples and reagents were transported, mixed, and washed in sequence on a paper chip for the detection of glucose, dopamine, and uric acid in human serum (Figure 19b–d).

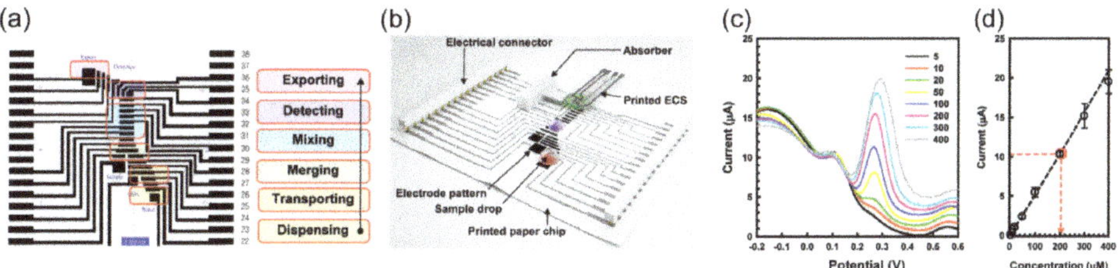

Figure 19. Electrochemical detection by using a p-DMF chip. (**a**) Programming sample delivery on the chip for assay protocols. (**b**) Schematic illustration of a p-DMF chip with an electrochemical sensor. (**c**) Electrochemical signal received from uric acid of various concentrations and (**d**) the calibration curve. Reprinted with permission from the authors of [28]. Copyright 2017 WILEY-VCH Verlag GmbH & Co. KGaA, Weinheim.

8. Challenges and Future Directions of Programmable Paper-Based Microfluidic Devices

Although paper-based microfluidic devices have made significant progress related to their fabrication method, operation technique, and application, they are still limited for analyzing challenging biomarkers such as cells and micro-nano bioparticles. Increasing the resolution of the p-CMF channels may allow such biomarkers to be studied by using p-CMF devices. Currently, because making very fine channels in p-CMF devices is more difficult than it is in polymer-based microfluidic devices, the latter devices are commonly used, instead of p-CMF devices, for the study of red blood cell deformation [109]. Moreover, being able to better control the characteristics of fluid transport such as the flow speed in p-CMF devices will pave the way for their application in a wide range of biomarker analyses. Unlike polymer-based microfluidic devices that are operated by external pumping systems, p-CMF devices are passively operated by capillary action. However, the flow of fluid in the p-CMF channel is weak and decreases with time. Because the surface of paper can be easily modified for chemical binding, p-CMF devices are commonly used for immobilizing enzymes, antibodies, and chemical reagents for chemical and biological assays [29].

Recently, the use of p-DMF devices for biomarker assays has increased significantly. With the recent development of fabrication methods, p-DMF devices can be fabricated easily and more affordably. Moreover, portable control systems and software for operation of the p-DMF devices are now well developed, so p-DMF devices are more flexible for fluid control. We encourage researchers to use p-DMF devices as microfluidic platforms for biomarker analyses.

9. Conclusions and Outlook

Our review on the recent development of programmable paper-based microfluidic devices shows rapid progress in the fundamental understanding and engineering of devices for advanced fluid sample handling. In the near future, programmable paper-based microfluidic devices should become promising tools for screening patients for a wide range of diseases. Because the devices can be programmed for better control of fluidic sample transport, delay, valving, mixing, and merging, more and more diseases can be detected using a single- or a multi-step assay protocol with highly sensitive signal detection. Even though these programmable paper-based microfluidic devices have notable advantages in term of fluidic handling, they are currently more expensive than traditional paper-based microfluidic devices because they require additional modifications involving the use of functional materials and/or advanced engineering. If the price of assay detection is to become more affordable, their fabrication cost must be reduced because these devices are disposable. We encourage researchers to continue making contributions on the development of paper-based microfluidic devices, especially contributions regarding cheaper fabrication and controllability methods, so that they can be used for a wide range of applications. The use of existing programmable paper-based microfluidic devices to explore new methods for detecting new biomarkers is strongly recommended.

Author Contributions: Figure preparation: V.S.; Resources: K.S.; Supervision: K.S.; Writing the original draft: V.S.; Writing, reviewing and editing the final manuscript: V.S., S.P., A.I.B., O.-S.K. and K.S.

Funding: The authors gratefully acknowledge support from the Basic Science Research Program (2018R1A6A1A03024940) through the National Research Foundation of Korea (NRF) funded by the Ministry of Science and ICT (MSIT) and from the Grant Program (20174010201150) funded by the Ministry of Trade, Industry & Energy, Korea. Grant Program (20174010201150) funded by the Ministry of Trade, Industry & Energy, Korea.

Conflicts of Interest: The authors declare no conflicts of interest.

References

1. Müller, R.H.; Clegg, D.L. Automatic Paper Chromatography. *Anal. Chem.* **1949**, *21*, 1123–1125. [CrossRef]
2. Martinez, A.W.; Phillips, S.T.; Butte, M.J.; Whitesides, G.M. Patterned paper as a platform for inexpensive, low-volume, portable bioassays. *Angew. Chem. Int. Ed. Engl.* **2007**, *46*, 1318–1320. [CrossRef] [PubMed]
3. Ko, H.; Lee, J.; Kim, Y.; Lee, B.; Jung, C.H.; Choi, J.H.; Kwon, O.S.; Shin, K. Active digital microfluidic paper chips with inkjet-printed patterned electrodes. *Adv. Mater.* **2014**, *26*, 2335–2340. [CrossRef] [PubMed]
4. Fobel, R.; Kirby, A.E.; Ng, A.H.; Farnood, R.R.; Wheeler, A.R. Paper microfluidics goes digital. *Adv. Mater.* **2014**, *26*, 2838–2843. [CrossRef] [PubMed]
5. Pollack, M.G.; Fair, R.B.; Shenderov, A.D. Electrowetting-based actuation of liquid droplets for microfluidic applications. *Appl. Phys. Lett.* **2000**, *77*, 1725–1726. [CrossRef]
6. Nelson, W.C.; Kim, C.-J.C. Droplet Actuation by Electrowetting-on-Dielectric (EWOD): A Review. *J. Adhes. Sci. Technol.* **2012**, *26*, 1747–1771. [CrossRef]
7. Moon, H.; Cho, S.K.; Garrell, R.L.; Kim, C.J. Low voltage electrowetting-on-dielectric. *J. Appl. Phys.* **2002**, *92*, 4080–4087. [CrossRef]
8. Monkkonen, L.; Edgar, J.S.; Winters, D.; Heron, S.R.; Mackay, C.L.; Masselon, C.D.; Stokes, A.A.; Langridge-Smith, P.R.R.; Goodlett, D.R. Screen-printed digital microfluidics combined with surface acoustic wave nebulization for hydrogen-deuterium exchange measurements. *J. Chromatogr. A* **2016**, *1439*, 161–166. [CrossRef]
9. Dixon, C.; Ng, A.H.; Fobel, R.; Miltenburg, M.B.; Wheeler, A.R. An inkjet printed, roll-coated digital microfluidic device for inexpensive, miniaturized diagnostic assays. *Lab Chip* **2016**, *16*, 4560–4568. [CrossRef]
10. Ng, A.H.C.; Fobel, R.; Fobel, C.; Lamanna, J.; Rackus, D.G.; Summers, A.; Dixon, C.; Dryden, M.D.M.; Lam, C.; Ho, M.; et al. A digital microfluidic system for serological immunoassays in remote settings. *Sci. Transl. Med.* **2018**, *10*, eaar6076. [CrossRef]
11. Sista, R.S.; Eckhardt, A.E.; Wang, T.; Graham, C.; Rouse, J.L.; Norton, S.M.; Srinivasan, V.; Pollack, M.G.; Tolun, A.A.; Bali, D.; et al. Digital Microfluidic Platform for Multiplexing Enzyme Assays: Implications for Lysosomal Storage Disease Screening in Newborns. *Clin. Chem.* **2011**, *57*, 1444. [CrossRef]
12. Sista, R.S.; Wang, T.; Wu, N.; Graham, C.; Eckhardt, A.; Winger, T.; Srinivasan, V.; Bali, D.; Millington, D.S.; Pamula, V.K. Multiplex newborn screening for Pompe, Fabry, Hunter, Gaucher, and Hurler diseases using a digital microfluidic platform. *Clin. Chim. Acta* **2013**, *424*, 12–18. [CrossRef]
13. Wan, L.; Chen, T.; Gao, J.; Dong, C.; Wong, A.H.-H.; Jia, Y.; Mak, P.-I.; Deng, C.-X.; Martins, R.P. A digital microfluidic system for loop-mediated isothermal amplification and sequence specific pathogen detection. *Sci. Rep.* **2017**, *7*, 14586. [CrossRef]
14. Sathyanarayanan, G.; Haapala, M.; Sikanen, T. Interfacing Digital Microfluidics with Ambient Mass Spectrometry Using SU-8 as Dielectric Layer. *Micromachines* **2018**, *9*. [CrossRef]
15. He, Y.; Wu, Y.; Fu, J.-Z.; Wu, W.-B. Fabrication of paper-based microfluidic analysis devices: A review. *RSC Adv.* **2015**, *5*, 78109–78127. [CrossRef]
16. Koesdjojo, M.T.; Pengpumkiat, S.; Wu, Y.; Boonloed, A.; Huynh, D.; Remcho, T.P.; Remcho, V.T. Cost Effective Paper-Based Colorimetric Microfluidic Devices and Mobile Phone Camera Readers for the Classroom. *J. Chem. Educ.* **2015**, *92*, 737–741. [CrossRef]
17. Tenda, K.; Ota, R.; Yamada, K.; Henares, G.T.; Suzuki, K.; Citterio, D. High-Resolution Microfluidic Paper-Based Analytical Devices for Sub-Microliter Sample Analysis. *Micromachines* **2016**, *7*. [CrossRef]
18. Ghaderinezhad, F.; Amin, R.; Temirel, M.; Yenilmez, B.; Wentworth, A.; Tasoglu, S. High-throughput rapid-prototyping of low-cost paper-based microfluidics. *Sci. Rep.* **2017**, *7*, 3553. [CrossRef]

19. Fu, E.; Lutz, B.; Kauffman, P.; Yager, P. Controlled reagent transport in disposable 2D paper networks. *Lab Chip* **2010**, *10*, 918–920. [CrossRef]

20. Mahmud, A.M.; Blondeel, J.E.; Kaddoura, M.; MacDonald, D.B. Features in Microfluidic Paper-Based Devices Made by Laser Cutting: How Small Can They Be? *Micromachines* **2018**, *9*. [CrossRef]

21. Fang, X.; Wei, S.; Kong, J. Paper-based microfluidics with high resolution, cut on a glass fiber membrane for bioassays. *Lab Chip* **2014**, *14*, 911–915. [CrossRef]

22. Song, M.-B.; Joung, H.-A.; Oh, Y.K.; Jung, K.; Ahn, Y.D.; Kim, M.-G. Tear-off patterning: A simple method for patterning nitrocellulose membranes to improve the performance of point-of-care diagnostic biosensors. *Lab Chip* **2015**, *15*, 3006–3012. [CrossRef]

23. Li, X.; Ballerini, D.R.; Shen, W. A perspective on paper-based microfluidics: Current status and future trends. *Biomicrofluidics* **2012**, *6*, 011301. [CrossRef]

24. Fu, E.; Downs, C. Progress in the development and integration of fluid flow control tools in paper microfluidics. *Lab Chip* **2017**, *17*, 614–628. [CrossRef]

25. Jeong, S.-G.; Kim, J.; Jin, S.H.; Park, K.-S.; Lee, C.-S. Flow control in paper-based microfluidic device for automatic multistep assays: A focused minireview. *Korean J. Chem. Eng.* **2016**, *33*, 2761–2770. [CrossRef]

26. Yang, Y.; Noviana, E.; Nguyen, M.P.; Geiss, B.J.; Dandy, D.S.; Henry, C.S. Paper-Based Microfluidic Devices: Emerging Themes and Applications. *Anal. Chem.* **2017**, *89*, 71–91. [CrossRef]

27. Tanev, G.; Madsen, J. A correct-by-construction design and programming approach for open paper-based digital microfluidics. In Proceedings of the 2017 Symposium on Design, Test, Integration and Packaging of MEMS/MOEMS (DTIP), Bordeaux, France, 29 May–1 June 2017; pp. 1–6.

28. Ruecha, N.; Lee, J.; Chae, H.; Cheong, H.; Soum, V.; Preechakasedkit, P.; Chailapakul, O.; Tanev, G.; Madsen, J.; Rodthongkum, N.; et al. Paper-Based Digital Microfluidic Chip for Multiple Electrochemical Assay Operated by a Wireless Portable Control System. *Adv. Mater. Technol.* **2017**, *2*, 1600267. [CrossRef]

29. Deng, J.Q.; Jiang, X.Y. Advances in Reagents Storage and Release in Self-Contained Point-of-Care Devices. *Adv. Mater. Technol.* **2019**, *4*, 1800625. [CrossRef]

30. Gong, M.M.; Sinton, D. Turning the Page: Advancing Paper-Based Microfluidics for Broad Diagnostic Application. *Chem. Rev.* **2017**, *117*, 8447–8480. [CrossRef]

31. Campbell, M.J.; Balhoff, B.J.; Landwehr, M.G.; Rahman, M.S.; Vaithiyanathan, M.; Melvin, T.A. Microfluidic and Paper-Based Devices for Disease Detection and Diagnostic Research. *Int. J. Mol. Sci.* **2018**, *19*. [CrossRef]

32. Lin, Y.; Gritsenko, D.; Feng, S.; Teh, Y.C.; Lu, X.; Xu, J. Detection of heavy metal by paper-based microfluidics. *Biosens. Bioelectron.* **2016**, *83*, 256–266. [CrossRef]

33. Samiei, E.; Tabrizian, M.; Hoorfar, M. A review of digital microfluidics as portable platforms for lab-on a-chip applications. *Lab Chip* **2016**, *16*, 2376–2396. [CrossRef]

34. Cate, D.M.; Adkins, J.A.; Mettakoonpitak, J.; Henry, C.S. Recent Developments in Paper-Based Microfluidic Devices. *Anal. Chem.* **2015**, *87*, 19–41. [CrossRef]

35. Yu, L.; Shi, Z.Z. Microfluidic paper-based analytical devices fabricated by low-cost photolithography and embossing of Parafilm (R). *Lab Chip* **2015**, *15*, 1642–1645. [CrossRef]

36. Mohammadi, S.; Maeki, M.; Mohamadi, R.M.; Ishida, A.; Tani, H.; Tokeshi, M. An instrument-free, screen-printed paper microfluidic device that enables bio and chemical sensing. *Analyst* **2015**, *140*, 6493–6499. [CrossRef]

37. Strong, E.B.; Schultz, S.A.; Martinez, A.W.; Martinez, N.W. Fabrication of Miniaturized Paper-Based Microfluidic Devices (MicroPADs). *Sci. Rep.* **2019**, *9*, 7. [CrossRef]

38. Lee, W.; Gomez, A.F. Experimental Analysis of Fabrication Parameters in the Development of Microfluidic Paper-Based Analytical Devices (μPADs). *Micromachines* **2017**, *8*. [CrossRef]

39. Gabriel, F.E.; Garcia, T.P.; Lopes, M.F.; Coltro, K.W. Paper-Based Colorimetric Biosensor for Tear Glucose Measurements. *Micromachines* **2017**, *8*. [CrossRef]

40. Juang, Y.J.; Li, W.S.; Chen, P.S. Fabrication of microfluidic paper-based analytical devices by filtration-assisted screen printing. *J. Taiwan Inst. Chem. Eng.* **2017**, *80*, 71–75. [CrossRef]

41. Mani, N.K.; Prabhu, A.; Biswas, S.K.; Chakraborty, S. Fabricating Paper Based Devices Using Correction Pens. *Sci. Rep.* **2019**, *9*, 1752. [CrossRef]

42. Li, X.; Tian, J.; Garnier, G.; Shen, W. Fabrication of paper-based microfluidic sensors by printing. *Colloids Surf. B Biointerfaces* **2010**, *76*, 564–570. [CrossRef]

43. Hamidon, N.N.; Hong, Y.M.; Salentijn, G.I.J.; Verpoorte, E. Water-based alkyl ketene dimer ink for user-friendly patterning in paper microfluidics. *Anal. Chim. Acta* **2018**, *1000*, 180–190. [CrossRef]

44. Xu, C.X.; Cai, L.F.; Zhong, M.H.; Zheng, S.Y. Low-cost and rapid prototyping of microfluidic paper-based analytical devices by inkjet printing of permanent marker ink. *RSC Adv.* **2015**, *5*, 4770–4773. [CrossRef]

45. Nie, J.F.; Zhang, Y.; Lin, L.W.; Zhou, C.B.; Li, S.H.; Zhang, L.M.; Li, J.P. Low-Cost Fabrication of Paper-Based Microfluidic Devices by One-Step Plotting. *Anal. Chem.* **2012**, *84*, 6331–6335. [CrossRef]

46. Lam, T.; Devadhasan, J.P.; Howse, R.; Kim, J. A Chemically Patterned Microfluidic Paper-based Analytical Device (C-μPAD) for Point-of-Care Diagnostics. *Sci. Rep.* **2017**, *7*, 1188. [CrossRef]

47. Nie, J.F.; Liang, Y.Z.; Zhang, Y.; Le, S.W.; Li, D.N.; Zhang, S.B. One-step patterning of hollow microstructures in paper by laser cutting to create microfluidic analytical devices. *Analyst* **2013**, *138*, 671–676. [CrossRef]

48. Thuo, M.M.; Martinez, R.V.; Lan, W.-J.; Liu, X.; Barber, J.; Atkinson, M.B.J.; Bandarage, D.; Bloch, J.-F.; Whitesides, G.M. Fabrication of Low-Cost Paper-Based Microfluidic Devices by Embossing or Cut-and-Stack Methods. *Chem. Mater.* **2014**, *26*, 4230–4237. [CrossRef]

49. Noh, H.; Phillips, S.T. Metering the Capillary-Driven Flow of Fluids in Paper-Based Microfluidic Devices. *Anal. Chem.* **2010**, *82*, 4181–4187. [CrossRef]

50. Ge, L.; Wang, S.M.; Song, X.R.; Ge, S.G.; Yu, J.H. 3D Origami-based multifunction-integrated immunodevice: Low-cost and multiplexed sandwich chemiluminescence immunoassay on microfluidic paper-based analytical device. *Lab Chip* **2012**, *12*, 3150–3158. [CrossRef]

51. Li, X.; Liu, X.Y. A Microfluidic Paper-Based Origami Nanobiosensor for Label-Free, Ultrasensitive Immunoassays. *Adv. Healthc. Mater.* **2016**, *5*, 1326–1335. [CrossRef]

52. Liu, H.; Crooks, R.M. Three-Dimensional Paper Microfluidic Devices Assembled Using the Principles of Origami. *J. Am. Chem. Soc.* **2011**, *133*, 17564–17566. [CrossRef]

53. Soum, V.; Cheong, H.; Kim, K.; Kim, Y.; Chuong, M.; Ryu, S.R.; Yuen, P.K.; Kwon, O.-S.; Shin, K. Programmable Contact Printing Using Ballpoint Pens with a Digital Plotter for Patterning Electrodes on Paper. *ACS Omega* **2018**, *3*, 16866–16873. [CrossRef]

54. He, Y.; Gao, Q.; Wu, W.-B.; Nie, J.; Fu, J.-Z. 3D Printed Paper-Based Microfluidic Analytical Devices. *Micromachines* **2016**, *7*. [CrossRef]

55. Juang, Y.-J.; Chen, P.-S.; Wang, Y. Rapid fabrication of microfluidic paper-based analytical devices by microembossing. *Sens. Actuators B Chem.* **2019**, *283*, 87–92. [CrossRef]

56. Postulka, N.; Striegel, A.; Krauße, M.; Mager, D.; Spiehl, D.; Meckel, T.; Worgull, M.; Biesalski, M. Combining Wax Printing with Hot Embossing for the Design of Geometrically Well-Defined Microfluidic Papers. *ACS Appl. Mater. Interfaces* **2019**, *11*, 4578–4587. [CrossRef]

57. Asano, H.; Shiraishi, Y. Development of paper-based microfluidic analytical device for iron assay using photomask printed with 3D printer for fabrication of hydrophilic and hydrophobic zones on paper by photolithography. *Anal. Chim. Acta* **2015**, *883*, 55–60. [CrossRef]

58. Songok, J.; Tuominen, M.; Teisala, H.; Haapanen, J.; Mäkelä, J.; Kuusipalo, J.; Toivakka, M. Paper-Based Microfluidics: Fabrication Technique and Dynamics of Capillary-Driven Surface Flow. *ACS Appl. Mater. Interfaces* **2014**, *6*, 20060–20066. [CrossRef]

59. Ghosh, A.; Ganguly, R.; Schutzius, T.M.; Megaridis, C.M. Wettability patterning for high-rate, pumpless fluid transport on open, non-planar microfluidic platforms. *Lab Chip* **2014**, *14*, 1538–1550. [CrossRef]

60. Elsharkawy, M.; Schutzius, T.M.; Megaridis, C.M. Inkjet patterned superhydrophobic paper for open-air surface microfluidic devices. *Lab Chip* **2014**, *14*, 1168–1175. [CrossRef]

61. Sones, C.L.; Katis, I.N.; He, P.J.W.; Mills, B.; Namiq, M.F.; Shardlow, P.; Ibsen, M.; Eason, R.W. Laser-induced photo-polymerisation for creation of paper-based fluidic devices. *Lab Chip* **2014**, *14*, 4567–4574. [CrossRef]

62. Zhang, Y.; Ren, T.; He, J. Inkjet Printing Enabled Controllable Paper Superhydrophobization and Its Applications. *ACS Appl. Mater. Interfaces* **2018**, *10*, 11343–11349. [CrossRef]

63. Zhang, Y.; Ren, T.; Li, T.; He, J.; Fang, D. Paper-Based Hydrophobic/Lipophobic Surface for Sensing Applications Involving Aggressive Liquids. *Adv. Mater. Interfaces* **2016**, *3*, 1600672. [CrossRef]

64. Bruzewicz, D.A.; Reches, M.; Whitesides, G.M. Low-Cost Printing of Poly(dimethylsiloxane) Barriers To Define Microchannels in Paper. *Anal. Chem.* **2008**, *80*, 3387–3392. [CrossRef]

65. Yafia, M.; Shukla, S.; Najjaran, H. Fabrication of digital microfluidic devices on flexible paper-based and rigid substrates via screen printing. *J. Micromech. Microeng.* **2015**, *25*, 057001. [CrossRef]

66. Abadian, A.; Sepehri Manesh, S.; Jafarabadi Ashtiani, S. Hybrid paper-based microfluidics: Combination of paper-based analytical device (µPAD) and digital microfluidics (DMF) on a single substrate. *Microfluid Nanofluidics* **2017**, *21*, 65. [CrossRef]

67. Abadian, A.; Jafarabadi-Ashtiani, S. Paper-based digital microfluidics. *Microfluid Nanofluidics* **2014**, *16*, 989–995. [CrossRef]

68. Jafry, A.T.; Lee, H.; Tenggara, A.P.; Lim, H.; Moon, Y.; Kim, S.H.; Lee, Y.; Kim, S.M.; Park, S.; Byun, D.; et al. Double-sided electrohydrodynamic jet printing of two-dimensional electrode array in paper-based digital microfluidics. *Sens. Actuators B Chem.* **2019**, *282*, 831–837. [CrossRef]

69. Soum, V.; Kim, Y.; Park, S.; Chuong, M.; Ryu, R.S.; Lee, H.S.; Tanev, G.; Madsen, J.; Kwon, O.-S.; Shin, K. Affordable Fabrication of Conductive Electrodes and Dielectric Films for a Paper-Based Digital Microfluidic Chip. *Micromachines* **2019**, *10*. [CrossRef]

70. Cheong, H.; Oh, H.; Kim, Y.; Kim, Y.; Soum, V.; Choi, J.H.; Kwon, O.S.; Shin, K. Effects of Silicone Oil on Electrowetting to Actuate a Digital Microfluidic Drop on Paper. *J. Nanosci. Nanotechnol.* **2018**, *18*, 7147–7150. [CrossRef]

71. Hong, S.; Kim, W. Dynamics of water imbibition through paper channels with wax boundaries. *Microfluid Nanofluidics* **2015**, *19*, 845–853. [CrossRef]

72. Walji, N.; MacDonald, D.B. Influence of Geometry and Surrounding Conditions on Fluid Flow in Paper-Based Devices. *Micromachines* **2016**, *7*. [CrossRef]

73. Adkins, J.A.; Noviana, E.; Henry, C.S. Development of a Quasi-Steady Flow Electrochemical Paper-Based Analytical Device. *Anal. Chem.* **2016**, *88*, 10639–10647. [CrossRef]

74. Renault, C.; Li, X.; Fosdick, S.E.; Crooks, R.M. Hollow-channel paper analytical devices. *Anal. Chem.* **2013**, *85*, 7976–7979. [CrossRef]

75. Channon, R.B.; Nguyen, M.P.; Scorzelli, A.G.; Henry, E.M.; Volckens, J.; Dandy, D.S.; Henry, C.S. Rapid flow in multilayer microfluidic paper-based analytical devices. *Lab Chip* **2018**, *18*, 793–802. [CrossRef]

76. Giokas, D.L.; Tsogas, G.Z.; Vlessidis, A.G. Programming fluid transport in paper-based microfluidic devices using razor-crafted open channels. *Anal. Chem.* **2014**, *86*, 6202–6207. [CrossRef]

77. Weng, C.-H.; Chen, M.-Y.; Shen, C.-H.; Yang, R.-J. Colored wax-printed timers for two-dimensional and three-dimensional assays on paper-based devices. *Biomicrofluidics* **2014**, *8*, 066502. [CrossRef]

78. Jang, I.; Song, S. Facile and precise flow control for a paper-based microfluidic device through varying paper permeability. *Lab Chip* **2015**, *15*, 3405–3412. [CrossRef]

79. Choi, J.R.; Yong, K.W.; Tang, R.; Gong, Y.; Wen, T.; Yang, H.; Li, A.; Chia, Y.C.; Pingguan-Murphy, B.; Xu, F. Lateral Flow Assay Based on Paper–Hydrogel Hybrid Material for Sensitive Point-of-Care Detection of Dengue Virus. *Adv. Healthc. Mater.* **2017**, *6*, 1600920. [CrossRef]

80. Chu, W.; Chen, Y.; Liu, W.; Zhao, M.; Li, H. Paper-based chemiluminescence immunodevice with temporal controls of reagent transport technique. *Sens. Actuators B Chem.* **2017**, *250*, 324–332. [CrossRef]

81. Lutz, B.; Liang, T.; Fu, E.; Ramachandran, S.; Kauffman, P.; Yager, P. Dissolvable fluidic time delays for programming multi-step assays in instrument-free paper diagnostics. *Lab Chip* **2013**, *13*, 2840–2847. [CrossRef]

82. Houghtaling, J.; Liang, T.; Thiessen, G.; Fu, E. Dissolvable Bridges for Manipulating Fluid Volumes in Paper Networks. *Anal. Chem.* **2013**, *85*, 11201–11204. [CrossRef]

83. Jahanshahi-Anbuhi, S.; Henry, A.; Leung, V.; Sicard, C.; Pennings, K.; Pelton, R.; Brennan, J.D.; Filipe, C.D.M. Paper-based microfluidics with an erodible polymeric bridge giving controlled release and timed flow shutoff. *Lab Chip* **2014**, *14*, 229–236. [CrossRef]

84. Rivas, L.; Medina-Sanchez, M.; de la Escosura-Muniz, A.; Merkoci, A. Improving sensitivity of gold nanoparticle-based lateral flow assays by using wax-printed pillars as delay barriers of microfluidics. *Lab Chip* **2014**, *14*, 4406–4414. [CrossRef]

85. Preechakasedkit, P.; Siangproh, W.; Khongchareonporn, N.; Ngamrojanavanich, N.; Chailapakul, O. Development of an automated wax-printed paper-based lateral flow device for alpha-fetoprotein enzyme-linked immunosorbent assay. *Biosens. Bioelectron.* **2018**, *102*, 27–32. [CrossRef]

86. Choi, J.R.; Liu, Z.; Hu, J.; Tang, R.; Gong, Y.; Feng, S.; Ren, H.; Wen, T.; Yang, H.; Qu, Z.; et al. Polydimethylsiloxane-Paper Hybrid Lateral Flow Assay for Highly Sensitive Point-of-Care Nucleic Acid Testing. *Anal. Chem.* **2016**, *88*, 6254–6264. [CrossRef]

87. He, P.J.W.; Katis, I.N.; Eason, R.W.; Sones, C.L. Engineering fluidic delays in paper-based devices using laser direct-writing. *Lab Chip* **2015**, *15*, 4054–4061. [CrossRef]

88. Toley, B.J.; McKenzie, B.; Liang, T.; Buser, J.R.; Yager, P.; Fu, E. Tunable-Delay Shunts for Paper Microfluidic Devices. *Anal. Chem.* **2013**, *85*, 11545–11552. [CrossRef]
89. Tang, R.; Yang, H.; Gong, Y.; Liu, Z.; Li, X.; Wen, T.; Qu, Z.; Zhang, S.; Mei, Q.; Xu, F. Improved Analytical Sensitivity of Lateral Flow Assay using Sponge for HBV Nucleic Acid Detection. *Sci. Rep.* **2017**, *7*, 1360. [CrossRef]
90. Akyazi, T.; Saez, J.; Elizalde, J.; Benito-Lopez, F. Fluidic flow delay by ionogel passive pumps in microfluidic paper-based analytical devices. *Sens. Actuators B Chem.* **2016**, *233*, 402–408. [CrossRef]
91. Songok, J.; Toivakka, M. Controlling capillary-driven surface flow on a paper-based microfluidic channel. *Microfluid Nanofluidics* **2016**, *20*, 63. [CrossRef]
92. Songok, J.; Toivakka, M. Enhancing Capillary-Driven Flow for Paper-Based Microfluidic Channels. *ACS Appl. Mater. Interfaces* **2016**, *8*, 30523–30530. [CrossRef]
93. Niedl, R.R.; Beta, C. Hydrogel-driven paper-based microfluidics. *Lab Chip* **2015**, *15*, 2452–2459. [CrossRef]
94. Wang, Q.; Li, Z.; Cheong, H.; Kwon, O.-S.; Yao, H.; Ho, T.-Y.; Shin, K.; Li, B.; Schlichtmann, U.; Cai, Y. Control-fluidic CoDesign for paper-based digital microfluidic biochips. In Proceedings of the 35th International Conference on Computer-Aided Design, Austin, TX, USA, 7–10 November 2016; pp. 1–8.
95. Li, J.; Wang, S.; Li, K.S.; Ho, T. Congestion- and timing-driven droplet routing for pin-constrained paper-based microfluidic biochips. In Proceedings of the 2016 21st Asia and South Pacific Design Automation Conference (ASP-DAC), Macao, China, 25–28 January 2016; pp. 593–598.
96. Sriram, G.; Bhat, M.P.; Patil, P.; Uthappa, U.T.; Jung, H.-Y.; Altalhi, T.; Kumeria, T.; Aminabhavi, T.M.; Pai, R.K.; Madhuprasad; et al. Paper-based microfluidic analytical devices for colorimetric detection of toxic ions: A review. *TrAC Trends Anal. Chem.* **2017**, *93*, 212–227. [CrossRef]
97. Busa, S.L.; Mohammadi, S.; Maeki, M.; Ishida, A.; Tani, H.; Tokeshi, M. Advances in Microfluidic Paper-Based Analytical Devices for Food and Water Analysis. *Micromachines* **2016**, *7*. [CrossRef]
98. Ye, H.; Xia, X. Enhancing the sensitivity of colorimetric lateral flow assay (CLFA) through signal amplification techniques. *J. Mater. Chem. B* **2018**, *6*, 7102–7111. [CrossRef]
99. Yang, W.; Li, X.-B.; Liu, G.-W.; Zhang, B.-B.; Zhang, Y.; Kong, T.; Tang, J.-J.; Li, D.-N.; Wang, Z. A colloidal gold probe-based silver enhancement immunochromatographic assay for the rapid detection of abrin-a. *Biosens. Bioelectron.* **2011**, *26*, 3710–3713. [CrossRef]
100. Anfossi, L.; Di Nardo, F.; Giovannoli, C.; Passini, C.; Baggiani, C. Increased sensitivity of lateral flow immunoassay for ochratoxin A through silver enhancement. *Anal. Bioanal. Chem.* **2013**, *405*, 9859–9867. [CrossRef]
101. Ren, W.; Cho, I.-H.; Zhou, Z.; Irudayaraj, J. Ultrasensitive detection of microbial cells using magnetic focus enhanced lateral flow sensors. *Chem. Commun.* **2016**, *52*, 4930–4933. [CrossRef]
102. Parolo, C.; de la Escosura-Muñiz, A.; Merkoçi, A. Enhanced lateral flow immunoassay using gold nanoparticles loaded with enzymes. *Biosens. Bioelectron.* **2013**, *40*, 412–416. [CrossRef]
103. Gao, Z.; Ye, H.; Tang, D.; Tao, J.; Habibi, S.; Minerick, A.; Tang, D.; Xia, X. Platinum-Decorated Gold Nanoparticles with Dual Functionalities for Ultrasensitive Colorimetric in Vitro Diagnostics. *Nano Lett.* **2017**, *17*, 5572–5579. [CrossRef]
104. Gabriel, E.F.M.; Garcia, P.T.; Cardoso, T.M.G.; Lopes, F.M.; Martins, F.T.; Coltro, W.K.T. Highly sensitive colorimetric detection of glucose and uric acid in biological fluids using chitosan-modified paper microfluidic devices. *Analyst* **2016**, *141*, 4749–4756. [CrossRef]
105. Katis, I.N.; He, P.J.W.; Eason, R.W.; Sones, C.L. Improved sensitivity and limit-of-detection of lateral flow devices using spatial constrictions of the flow-path. *Biosens. Bioelectron.* **2018**, *113*, 95–100. [CrossRef]
106. Jahanshahi-Anbuhi, S.; Kannan, B.; Pennings, K.; Monsur Ali, M.; Leung, V.; Giang, K.; Wang, J.; White, D.; Li, Y.; Pelton, R.H.; et al. Automating multi-step paper-based assays using integrated layering of reagents. *Lab Chip* **2017**, *17*, 943–950. [CrossRef]
107. Wang, H.; Chen, L.G.; Sun, L.N. Digital microfluidics: A promising technique for biochemical applications. *Front. Mech. Eng. Prc.* **2017**, *12*, 510–525. [CrossRef]

108. Jang, I.; Ko, H.; You, G.; Lee, H.; Paek, S.; Chae, H.; Lee, J.H.; Choi, S.; Kwon, O.-S.; Shin, K.; et al. Application of paper EWOD (electrowetting-on-dielectrics) chip: Protein tryptic digestion and its detection using MALDI-TOF mass spectrometry. *BioChip J.* **2017**, *11*, 146–152. [CrossRef]

109. Bento, D.; Rodrigues, O.R.; Faustino, V.; Pinho, D.; Fernandes, S.C.; Pereira, I.A.; Garcia, V.; Miranda, M.J.; Lima, R. Deformation of Red Blood Cells, Air Bubbles, and Droplets in Microfluidic Devices: Flow Visualizations and Measurements. *Micromachines* **2018**, *9*. [CrossRef]

 *micromachines*

Article

Tangential Flow Microfiltration for Viral Separation and Concentration

Yi Wang [1], Keely Keller [2] and Xuanhong Cheng [1,2,*

[1]   Department of Materials Science and Engineering, Lehigh University, Bethlehem, PA 18015, USA; yiw716@lehigh.edu

[2]   Department of Bioengineering, Lehigh University, Bethlehem, PA 18015, USA; keely.ann.keller@gmail.com

*   Correspondence: xuc207@lehigh.edu

Received: 26 April 2019; Accepted: 9 May 2019; Published: 12 May 2019

Abstract: Microfluidic devices that allow biological particle separation and concentration have found wide applications in medical diagnosis. Here we present a viral separation polydimethylsiloxane (PDMS) device that combines tangential flow microfiltration and affinity capture to enrich HIV virus in a single flow-through fashion. The set-up contains a filtration device and a tandem resistance channel. The filtration device consists of two parallel flow channels separated by a polycarbonate nanoporous membrane. The resistance channel, with dimensions design-guided by COMSOL simulation, controls flow permeation through the membrane in the filtration device. A flow-dependent viral capture efficiency is observed, which likely reflects the interplay of several processes, including specific binding of target virus, physical deposition of non-specific particles, and membrane cleaning by shear flow. At the optimal flow rate, nearly 100% of viral particles in the permeate are captured on the membrane with various input viral concentrations. With its easy operation and consistent performance, this microfluidic device provides a potential solution for HIV sample preparation in resource-limited settings.

Keywords: HIV diagnostics; cross-flow filtration; microfluidic device; COMSOL; nanoporous membrane

---

1. Introduction

Diagnosis of viral infection such as HIV is still largely limited by resources in high-incidence areas such as sub-Saharan Africa. Microfluidic devices have been widely explored to tackle this challenge as they offer the potential of miniaturized, portable devices that are easy to operate and provide results faster than traditional viral-load or immunoassay tests [1–3]. In terms of measuring the HIV concentration in circulation, i.e., the viral load, challenges exist to separate and enrich viral particles from a complex fluid that contains many non-specific particles in the similar size range. For example, after primary HIV infection, the viral load could raise to thousands of particles per microliter of blood while there could easily be billions of extracellular vesicles in the same volume [4,5]. It has been an active research topic to efficiently extract viral particles of interest with simple procedures to facilitate subsequent viral detection.

Various microfluidic devices have been fabricated to separate, sort and concentrate biological particles from complex fluids [6–8]. Paper-based microfluidic devices have been introduced for point-of-care testing especially in resource-limited settings, as they are inexpensive to fabricate and allow passive transport of fluids without active pumping [9–12]. PDMS-based microfluidic devices, on the other hand, have improved optical, thermal and mechanical properties compared to paper, glass or silicon-based 'lab-on-a-chip' devices [1,13–15]. Depending on the mechanisms, the methods can be divided into physical and affinity separation approaches. Different physical properties, such as particle size, density, shape, electrical and dielectrical properties have been employed to separate viruses from other species. While they are flexible and have high throughputs, electrical separations often require special solutions to suspend the sample to reduce joule heating and/or electrical field shielding. Acoustic and optical separations also face the challenge of heating, especially in separating sub-micron and nanometer-sized species. Hydrodynamic separations, including pinched, inertia and dean flow fractionation are often not effective to separate nanoparticles. Dead-end filtration, although easy to operate, is prone to clogging. Furthermore, these physical methods generally lack specificity, as the target particles usually share similar physical properties to some of the non-specific species [6,8,16–18]. Affinity separation uses a solid matrix decorated with a receptor to specifically bind the target particles. For example, magnetic separation uses affinity magnetic beads combined with a magnetic field to temporarily immobilize the bound species while washing away all the other species. Although easy to operate and versatile towards targets of different sizes, magnetic separation could face variable yields especially at low target concentrations [19–21]. Flatbed microfluidic channels coated with antibodies have low capture efficiency for viruses due to the size mismatch between the channel and target particles. Introducing micro- and nanostructures such as posts and pores into microchannels or using nanofludics directly could significantly improve the interactions and capture yield, whereas they face issues such as fabrication complexity, high flow resistance and low throughput [22,23].

To address the challenges discussed above, especially the lack of specificity in physical separation and low throughput in affinity isolation of virus, we propose the combination of tangential flow microfiltration and affinity capture to separate HIV viruses. Nanofiltration is widely used in the pharmaceutical industry to eliminate virus contamination in blood-based products, using membranes with cut-off sizes smaller than viruses [24,25]. To improve throughput and reduce cake formation, one typical approach is cross-flow filtration, also referred to as tangential filtration. The fundamental idea is to force solute through a membrane while maintaining a tangential flow to clean the membrane. Compared to dead-end filtration, cross-flow filtration is able to achieve high separation efficiency while maintaining a high throughput by reducing membrane clogging. Beyond the conventional use to eliminate undesirable virus contamination, cross-flow filtration has been applied to harvest viruses and cells in microfluidic devices [26–31]. While cross-flow microfiltration separates species based on the physical size alone, functionalizing the membrane with affinity chemistry renders it specifically towards the surface biochemistry of the target species. Combining the tangential microfiltration and immunoaffinity capture methods in a microfluidic channel, Mittal et al. demonstrated isolation of rare cancer cells with a capture efficiency of 70% [32]. However, to our knowledge, the approach has not been evaluated for virus separation.

In this work, we utilize a tangential flow polydimethylsiloxane (PDMS) device with a sandwiched nanoporous membrane to isolate and concentrate HIV virus. Cross-flow filtration is coupled with affinity separation to promote specificity and efficiency of HIV capture. As shown in Figure 1b, by applying an external resistance channel to control the permeation volume, the viral sample is partially pushed towards the membrane in the top channel and partially moves tangential to the membrane. Such diverted flow promotes the interaction between viral particles and the functional membrane surface. During filtration, particles smaller than the pores are able to cross the membrane and exit via the bottom channel outlet (permeate) while large particles in the sample exit through the top outlet (retentate). This easy-to-operate approach offers efficient viral sample processing with high throughput.

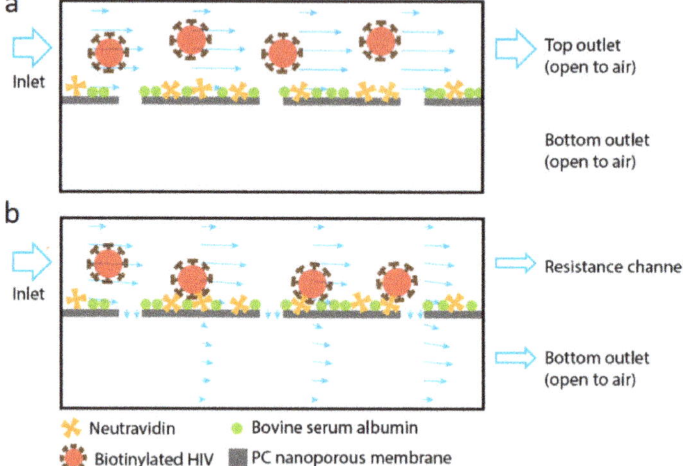

Figure 1. Schematics of the working principle of the device used in this work. The device is comprised of two flat channels separated vertically by a polycarbonate (PC) nanoporous membrane (pore size of 50 nm). The membrane is functionalized with NeutrAvidin to capture the biotinylated virus, with nonspecific binding blocked by bovine serum albumin. (**a**) When there is no cross-membrane flow, the virus primarily flows through the top channel. (**b**) When filtration is promoted using a resistance channel connected to the top channel, fluid enters the device through the top channel inlet and exits from both top (retentate) and bottom (permeate) outlets of the filtration device. The virus is pushed onto the membrane and captured by affinity binding. The arrows in the channels indicate local fluid velocity magnitude and direction.

2. Materials and Methods

2.1. Materials

SU-8 photoresist was purchased from MicroChem (Newton, MA, USA). 3″ silicon wafers were purchased from Silicon Inc. (Boise, ID, USA). Sylgard 184 silicone elastomer kit was obtained from Dow Corning (Midland, MI, USA). Nuclepore track-etched polycarbonate (PC) membrane (pore size of 50 nm) was purchased from Thomas Scientific (Swedesboro, NJ, USA). Toluene and lyophilized bovine serum albumin (BSA) were obtained from Sigma-Aldrich (St. Louis, MO, USA). Phosphate buffered saline (PBS) was obtained from Mediatech (Herndon, VA, USA). NeutrAvidin and triton-X 100 were purchased from Thermo Scientific (Rockford, IL, USA). An HIV-1 p24 enzyme-linked immunosorbent assay (ELISA) kit was obtained from PerkinElmer (Waltham, MA, USA).

2.2. Assembly of Porous Membrane and Main Channels

As shown in Figure 2, the filtration device contains a top channel and a bottom channel separated by a track-etched PC membrane with 50 nm pores. The height and length of two channels are identical of 0.1 mm and 30 mm while the top channel is 3 mm in width and 1 mm wider than the bottom channel for ease of channel alignment during device assembly. The bottom channel also contains cylinder posts 0.5 mm in diameter to support the membrane from sagging. Post positions are shown in the inset in Figure 2a.

Both the top and bottom PDMS channels were fabricated by standard soft lithography techniques. Briefly, SU-8 photoresist was patterned on silicon wafers in the clean room using standard photolithography. Afterwards, PDMS base and curing agent (Sylgard 184 silicone elastomer kit) were mixed at a 10:1 ratio and poured on the wafer molds. After 30-min degassing, the pre-polymer mixture was baked at 60 °C overnight to fully cure the PDMS. The cured PDMS was then removed from the molds, resulting in an open top channel and a post-containing bottom channel. The inlet and outlet ports of each channel were punched using a blunt-tip needle.

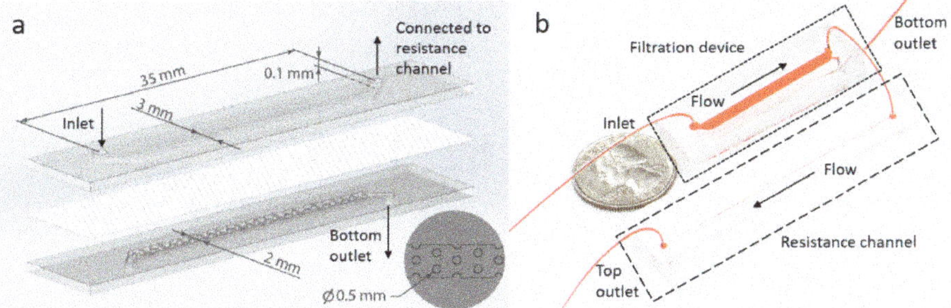

Figure 2. Schematic and photograph of the device. (**a**) A schematic showing the filtration device. A nanoporous PC membrane (50 nm pores) was assembled between two PDMS slabs containing molded microfluidic channels. The inlet in the top channel and outlets of both channels are labeled together with channel dimensions. The bottom right inset shows a top view of the molded PDMS posts which are regularly distributed in the bottom channel to physically support the membrane. (**b**) Photograph of the filtration device connected to a 65 mm long resistance channel. The devices are filled with food dye for visualization.

Next, a 50-nm pore-size PC membrane with a surface area comparable to the PDMS slab/channel was adhered between the two channels using a PDMS 'glue' which is a mixture of PDMS pre-polymer and toluene in a 1:1 ratio (w/w) [33]. 0.5 mL of the glue mixture was spin-coated on a glass slide (25 mm × 75 mm) for 1 min at 2000 rpm. As soon as spinning was finished, PDMS slabs/channels were placed on the glass slide with the channel side facing down for 30 s. The PC membrane was placed on top of the glue-coated bottom channel, and then the glue-coated top channel PDMS layer was aligned and put on top of the membrane and bottom channel. The thinly spin-coated layer of glue prevented glue from squeezing into or occluding the PDMS channels during assembly. Assembled devices were baked at 60 °C for 5 h to fully cure the PDMS glue. Pieces of tubing (10 cm long) were inserted into the inlet and outlet ports.

2.3. Resistance Channel and Flow Tests

A thin and straight microchannel was connected with the top outlet of the filtration device to control the flow resistance in the tangential flow path. The resistance channel is 0.2 mm in width, 0.04 mm in height and 65 mm in length. The procedure to fabricate the resistance channel in PDMS using standard soft lithography is the same as that described above for the filtration channels. The cured PDMS resistance channel was bonded to a glass slide (75 mm × 25 mm) using oxygen plasma (Nordson MARCH, Concord, CA, USA).

Flow tests were performed with and without the resistance channel connected to the top outlet (retentate) of the filtration device. 250 µL PBS was introduced into the top inlet of the filtration devices at different flow rates using a syringe pump (Chemyx, Stafford, TX, USA). Outflow was collected from both the retentate and permeate outlets, and weighed to calculate the volume ratio of solution draining through the membrane.

2.4. Membrane Functionalization and Viral Capture Test

Before the viral capture test, 250 µL deionized water was first injected by hand through the top inlet of the filtration device followed by 250 µL PBS buffer injection. Then, 150 µL NeutrAvidin at a concentration of 1 mg/mL in 1% BSA/PBS (w/v) was injected and incubated at room temperature for 2 h for membrane functionalization. Afterwards, PBS was hand injected to remove loosely bound

NeutrAvidin from the channel. Next, the device was filled with 1% BSA/PBS solution and incubated at room temperature for 30 min in order to block any bare uncoated membrane. PBS buffer was used to wash the channel and purge air bubbles afterwards. At each incubation step, the tubing was clamped to avoid any bubbles being introduced into the device. No resistance channel was used in the membrane functionalization procedure. After membrane functionalization, the resistance channel was connected to the top outlet of the filtration device to add flow resistance to the tangential flow path.

Pseudotyped HIV was cultured [34] and biotinylated [35] as reported before, and diluted to final concentrations using 1% BSA/PBS. 250 μL viral solution was pumped from the top inlet at defined flow rates. After this virus capture step, the resistance channel was disconnected and PBS was used to flush away unbound virus. Subsequently, 250 μL 0.5% Triton X-100 was injected at 2400 μL/h flow rate using a syringe pump from the top inlet to lyse the captured virus. Outflow from the top channel outlet was collected. HIV concentration in the lysate was determined using a commercial p24 ELISA assay according to the manufacturer's recommended procedures. The HIV concentration in the lysate was compared to the input HIV concentration in order to calculate the virus capture efficiency.

2.5. Fluid Dynamics Simulation

3D simulation was carried out by COMSOL Multiphysics software (version 5.3a, COMSOL Inc., Stockholm, Sweden) to optimize fluid flow in the microfluidic channel with a sandwiched membrane. The simulated dimensions of the filtration device match those used experimentally. Stabilizing posts are not included in the simulation. Water was selected as the fluid in the channel domains. The membrane domain was set as a 10-μm-thick porous matrix with 10% porosity and permeability of 2.02×10^{-18} m^2, which were obtained from the manufacturer's data sheet and independent filtration tests, respectively. The flow in both the top and bottom channels was assumed to be steady state laminar flow and was described by the Navier–Stokes equation, while the membrane region was characterized by a Forchheimer-corrected version of the Brinkman equation [36]. A user-controlled octahedron mesh was applied to all the domains. Mesh convergence was confirmed. A no-slip boundary condition was applied to all solid walls.

3. Results

Combining cross-flow microfiltration and affinity capture, we hypothesize that a high capture efficiency of viral particles can be achieved with continuous flow in a microfluidic channel. The filtration device consists of two parallel channels separated by a permeable PC nanoporous membrane. The viral sample flows in from the inlet at the top channel whereas two outlets are available at the other end of both the top and bottom channels. Thus, the sample can flow parallel to the membrane out of the top outlet as the retentate, or cross the membrane and exit from the bottom outlet as the permeate. The cross-membrane flow carries bioparticles from the top channel towards the membrane for their capture, while the tangential flow above the membrane continuously cleans the membrane from clogging. Specific viral capture is achieved through pre-functionalization of the membrane. Here, we study how the viral capture is influenced by top channel flow rates and viral concentrations.

3.1. Computational Analysis of Fluid Permeation through the Membrane

We first used COMSOL simulation to examine fluid flow in the proposed filtration device. To control the fraction of fluid exiting from the two channels separated by the membrane, an external straight microchannel is connected to the top retentate outlet of the filtration device, which increases the hydrodynamic resistance of the flow path tangential to the membrane. Without it, the membrane resistance is so high that little sample drains to the bottom permeate channel. As the resistance channel has a much smaller cross section compared to the top channel in the filtration device, it has much greater resistance, thus dominates the total resistance of the tangential flow path. In the membrane permeation flow path, the flow resistance is dominated by the porous membrane. Thus, by changing the length of the resistance channel, the permeation volume through the membrane can be readily controlled.

To guide the design of the resistance channel, the same dimension of the filtration device was used in the simulation as the experimental device (Figure 2a), and a permeable membrane separated the top and bottom channels. The resistance channel was simulated as an extension of the top channel beyond the filtration region (top of Figure 3a). It was 0.04 mm in depth and 0.2 mm in width, and with varying length from 45 to 85 mm. The two outlets, one at the bottom filtration channel and the other at the end of the resistance channel, were set to atmospheric pressure. The normal flow velocity at the inlet (at the left end of the top channel) was fixed at 2.4 mm/s in the simulation, which corresponded to a volumetric flow rate of 2400 µL/h.

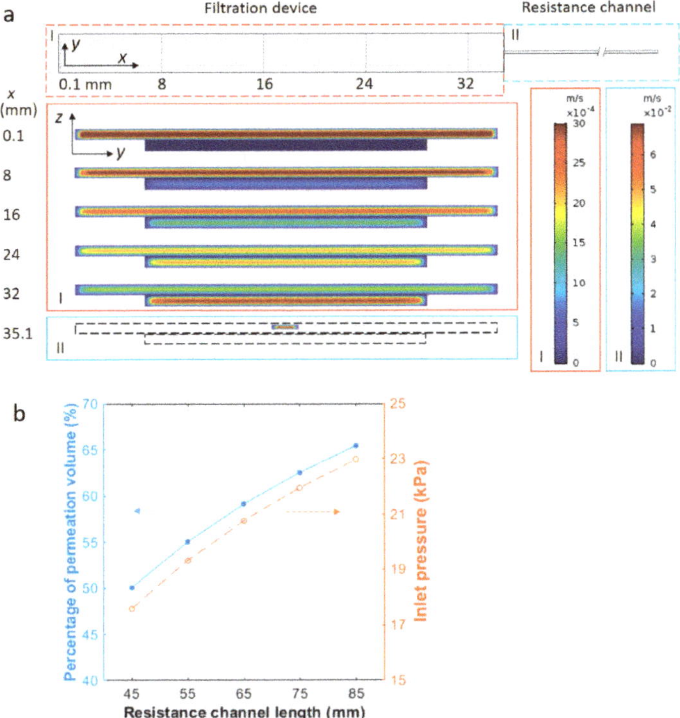

Figure 3. Magnitude profile of $+x$ velocity and permeation percentage in the device from COMSOL simulation. (**a**) The top view of the simulated device and profiles of the axial velocity in y-z cross sections of the channel at different distances from the inlet. The simulated geometry contains a filtration device (region I) of two channels separated by a porous membrane and a 65-mm-long resistance channel (region II). The top channel of the filtration device is connected to the resistance channel, while the bottom channel is open to air at the outlet. The velocity profiles correspond to positions of different cross sections along the $+x$ direction. These positions are labeled on the left of the velocity profiles in values (mm) and also as dashed lines in the x-y view of the simulated device (top). Axial flow is along the $+x$ direction and the average inlet velocity is 2.4 mm/s. The left color legend of velocity magnitudes corresponds to the filtration device (region I) and right color legend corresponds to the resistance channel (region II). The magnitude of the axial velocity is observed to decrease in the top channel, but increase gradually in the bottom channel from the inlet to the outlet. (**b**) Fluid fraction permeating through the membrane and inlet pressure as a function of the resistance channel length. The blue solid circles correspond to the left y-axis and empty orange circles correspond to the right y-axis. The lines are to guide the eyes. The cross-sectional area of the resistance channel is constant. The average flow velocity at the inlet is 2.4 mm/s in all cases.

An example of the simulation results is shown in Figure 3a. The resistance channel is 65 mm long in this case. Enlarged y-z cross sections in Figure 3a exhibit the x-component velocity profile at different locations from the inlet along the axial direction ($+x$ direction). As demonstrated by the velocity profile in various cross sections normal to the axial flow, the fluid velocity gradually decreases in the top channel and increases in the bottom permeate channel. This indicates that the fluid from the inlet gradually drains through the porous membrane and flows out from outlets above and below the membrane. The volumetric flow rates, obtained by surface integration of the axial velocity, yield 980 µL/h and 1420 µL/h at the top and bottom outlets respectively, corresponding to a fraction of permeation of 59.10%. The inlet pressure is around 20.8 kPa from the simulation. The average axial velocity in the bottom channel exceeds that of the top channel towards the end of the filtration device due partially to the greater volumetric flow rate and partially to the smaller cross-sectional area of the bottom channel. When the permeation fraction was evaluated for different inlet velocities of 0.3, 1.0, 1.7 and 3.1 mm/s, the value was found to be consistently around 59.10% and independent of the sample flow rate. This is understandable, as the ratio of the volumetric flow rate in the two parallel flow paths is equal to the inverse ratio of the flow resistance, and is independent of the input flow rate. The inlet pressure was also evaluated from the COMSOL simulation and was found to increase linearly with the average input velocity, changing from 2.6 kPa at 0.3 mm/s to 26.8 kPa at 3.1 mm/s. All of these pressure values are well below the burst pressure of the sandwiched device, which was found to be 70–100 kPa using a homemade pressure-testing device.

The fraction of permeated fluid volume was then simulated with various resistance channel lengths. Using the same cross-sectional area of 0.04 mm × 0.2 mm, channels of 45, 55, 65, 75 and 85 mm in length were tested at 2.4 mm/s inlet flow velocity. As the hydrodynamic resistance of a rectangular channel is proportional to its length, the longer resistance channels are expected to divert a larger amount of fluid to the bottom channel through the membrane. Results in Figure 3b show that the ratio of permeation volume rises from 50.02% with a 45 mm resistance channel to 65.38% with an 85 mm one. The inlet pressure increased from around 17.6 to 23.0 kPa, accordingly, due to greater flow resistance. The results indicate that the permeation volume can be easily controlled using properly selected resistance channels.

3.2. Experimental Evaluation of Fluid Permeation

Clogging is one of the main obstacles in microfiltration-based separation devices. Pressure accumulation caused by membrane fouling and clogging often lead to leakage and failure of the filtration device. In a tangential filtration device, however, the tangential flow clears the membrane fouling continuously, leading to greater permeate rate and extension of the device lifetime. As described above, the maximum backpressure and permeation fraction are controllable by the resistance channel design. In addition, sealing of the membrane in between the microchannels was also optimized to improve device integrity. To seal the membrane between two channels, we used a mortar layer of PDMS prepolymer as the glue between the PDMS slabs and the PC membrane [33]. A thin layer of PDMS precursor was spin-coated on the surface of a glass slide. The thickness of PDMS layer was controlled by the spinning speed and time [37]. Inked on the top and bottom PDMS blocks/channels, the precursor was able to penetrate through the porous membrane and hold on to the PDMS block/channel on the other side. After curing the glue layer at 60 °C, the membrane was sealed firmly between the two channels without any bubbles or membrane wrinkles. This method allows a uniformly thin layer of 'glue' to be transferred to the PDMS while accurately maintaining the channel geometry. Using a homemade pressure-testing device, the microfluidic device was found to be capable of holding up to 70–100 kPa of pressure. The maximum pressure in the experiments was designed to be less than 30 kPa through proper resistance channel selection to ensure integrity of the filtration device.

Based on the analysis above, a 65-mm-length resistance channel was used to collect up to 60% filtrate. Flow tests were then carried out experimentally to examine the fluid permeation ratio at various flow rates and inlet viral concentrations. Permeation was first tested at different flow rates with

PBS buffer and the results are shown in Figure 4a. As predicted, all fluid flows out of the top outlet without the resistance channel and no permeate is collectable from the bottom outlet (thus not plotted). With the 65 mm resistance channel, on the other hand, 60.54 ± 7.79%, 59.71 ± 5.78%, 57.95 ± 5.83%, 54.04 ± 3.82% and 56.94 ± 4.05% of the input fluid permeates through the membrane and flows out from the bottom outlet at respective flow rates of 300, 1000, 1700, 2400 and 3100 µL/h. No statistical difference is observed among the results at different flow rates. On average, the experimental permeation is 57.83 ± 2.56%, which compares well to the COMSOL simulation results of 59.10% permeation.

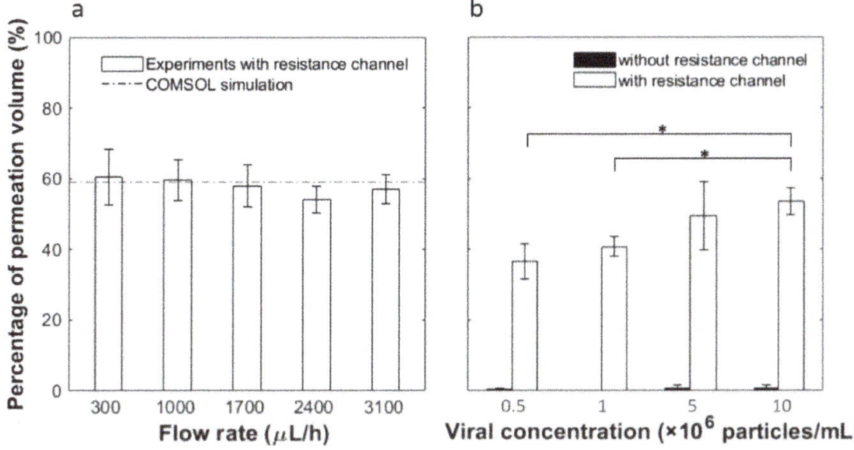

Figure 4. Fluid fraction permeating through the membrane as a function of (**a**) different flow rates and (**b**) inlet viral concentrations. In (**b**) the white bars are results from filtration devices connected with 65-mm-length resistance channels, while black bars are results from devices without. Error bars represent the standard deviation from at least 3 independent tests under the same condition. * indicates statistical difference based on two-tailed Student's t-test at a 95% confidence interval.

Next, we flowed solutions containing pseudotyped HIV at various concentrations through the microfluidic device at the same inlet flow rate of 2400 µL/h. As shown in Figure 4b, without the resistance channel, outflow from the bottom outlet is less than 1% in all cases. With the resistance channel, the permeation volumes are 36.56 ± 4.86%, 40.75 ± 2.81%, 49.49 ± 9.68% and 53.54 ± 3.77% of the feed at respective input concentrations of 5×10^5, 1×10^6, 5×10^6 and 1×10^7 particles/mL. Thus, a greater input particle concentration yields greater fluid permeation. This is counter-intuitive as one would expect that more viral particles could potentially block more membrane pores, leading to less permeation. However, this phenomenon may be explained by the carrier solution used in this work. Since HIV was spiked in PBS containing 1% BSA and the membrane pore size is 50 nm, BSA aggregates could clog the pores. When a greater virus concentration was used, the sample would contain less carrier solution or reduced BSA interference, thus allowing greater permeation.

3.3. Viral Capture Efficiency

Viral capture tests were performed with biotinylated pseudotyped HIV in the tangential flow device with a pre-functionalized membrane. First, samples at a concentration of 5×10^6 particles/mL were tested at different flow rates of 300, 1000, 1700, 2400 and 3100 µL/h. The resistance channel was disconnected from the device after applying the viral sample and the devices were rinsed with PBS at the same flow rate as the sample flow. Afterwards, Triton X-100 was flowed in to lyse the captured virus and release p24 antigen, the protein that makes up the core of HIV. Without the resistance channel, the lysis solution flowed almost exclusively out of the top outlet. Concentrations of p24 in the lysis solution and in the earlier flow-through solution were tested by a commercial ELISA assay. Mass conservation was confirmed by comparing the sum of the virus in different output fractions to that in the input. Capture efficiency was then calculated by the ratio of the p24 amount lysed out from the top channel to that in the input solution. As shown in Figure 5a, the capture efficiency is greatly dependent on the flow rate. A peak capture efficiency was found at flow rates of 1700–2400 µL/h. Viral capture percentage is 31.93 ± 8.25% at 1700 µL/h and 46.61 ± 16.19% at 2400 µL/h without statistical difference between these two flow conditions (95% confidence interval with Student's t-test). At flow rates of 300, 1000 and 3100 µL/h, the capture efficiency further deceases to 4%–15%, and the difference in capture efficiency is significant between these flow rates and 2400 µL/h. Possible causes of this viral capture flow rate dependence are discussed later in the text. At each flow rate, control experiments were also carried out using filtration devices without resistance channels. In such cases, permeation through the membrane was negligible, as was the binding efficiency. The low capture efficiency without the resistance channel is understandable by comparing the characteristic time of different transport processes. The residence time of samples in the filtration microchannel is around 10–120 s at the flow rates tested, while the diffusion time of 100 nm diameter particles across a 0.05-mm-high channel is much greater, at about 250 s. Thus, when filtration is restricted, diffusion of viral particles to the channel walls is limited, resulting in low capture efficiency.

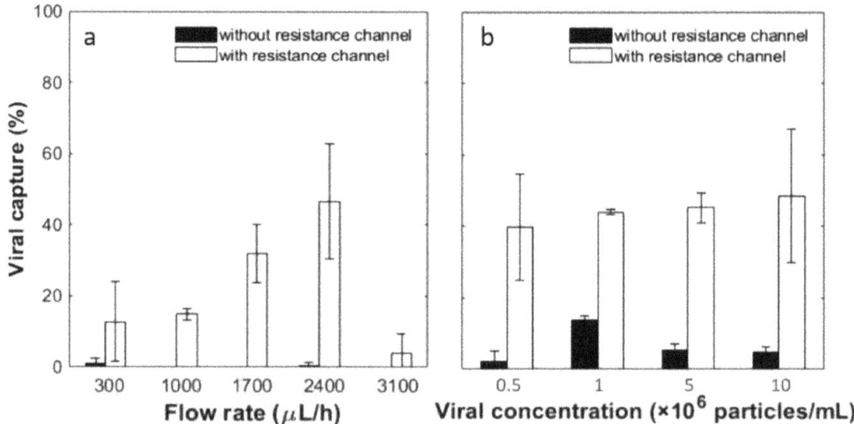

Figure 5. Viral capture yield as a function of (**a**) sample flow rates and (**b**) viral concentrations. Viral concentration used in (**a**) was 5×10^6 particles/mL. The flow rate applied in (**b**) was 2400 µL/h. The white bars are results from filtration devices connected with the 65 mm resistance channel, while black bars are results from devices without. Error bars represent the standard deviation from at least 3 devices under the same conditions. Statistical analysis was conducted for (**b**) by applying an independent Student's t-test at a 95% confidence interval and no statistically significant difference is observed among the white bars.

Next, a range of viral concentrations, 5×10^5, 1×10^6, 5×10^6 and 1×10^7 particles/mL, were tested as the input samples. The flow rate was fixed at 2400 µL/h, which showed the highest average yield at the concentration of 5×10^6 particles/mL as described above. With the resistance channel, 39.68 ± 14.84%, 43.99 ± 0.67%, 45.21 ± 4.18% and 48.54 ± 18.70% of viral particles are captured from the input of the lowest to highest viral concentrations. The capture efficiency was further divided by the percentage of permeation volume to gain an understanding of capture from the filtered fraction. This yields capture efficiencies of 106.79 ± 26.39%, 108.60 ± 8.91%, 92.70 ± 11.40% and 90.51 ± 18.70%, respectively, in the permeated fraction, and there is no statistical difference among the different concentration groups (95% confidence interval with Student's t-test). Without the resistance channel, the capture yields are significantly lower. From the input samples of the lowest to highest viral concentrations, 2.33 ± 2.71%, 13.95 ± 1.10%, 5.35 ± 1.80% and 4.68 ± 1.54% viral particles are captured. As little flow drained through the membrane without an external resistance channel according to COMSOL simulations and flow experiments discussed above, these yields indicate viral separation achieved by affinity capture only. This may explain capture efficiencies higher than 100% with resistance channels, after captures were normalized to permeation volume. Significant differences are observed between each pair of tests with and without the resistance channel at the same viral concentration. The results confirm that the tandem filtration device and resistance channel yield consistent and close to 100% viral capture from the membrane permeate in a wide range of input concentrations.

4. Discussion

Overall, we implemented tangential flow filtration in microfluidic devices for viral capture in a high throughput and simple flow-through fashion. The device contains a filtration device coupled with a 65-mm-long resistance channel to achieve a 6:4 split in volume of the permeate versus retentate. While maintaining laminar flows in both the permeate and retentate compartments, the filtration process promotes viral interactions with the functionalized capture membrane, while the tangential flow continuously clears non-specific deposits on the membrane. Compared to the filtration device alone without the resistance channel, which could represent a flatbed capture channel (as no fluid permeates through the membrane), the tangential filtration device allows significantly improved capture yields. Flow rate-dependent yields were observed with a peak at 1700–2400 μL/h. Reduced capture at slow flow is likely a result of competition between specific capture and non-specific deposit on the membrane. Similar competition has been reported before for immunoaffinity cell capture from blood. For example, capture of CD4+ T lymphocytes from blood in a flatbed microfluidic channel is more efficient when the near-wall shear is strong enough to clear the surface of red blood cells [38]. In our work, although a cell-free sample of unpurified virus culture spiked in a BSA solution is used, the presence of extracellular vesicles and protein aggregates in the virus culture and carrier solution are also capable of interacting nonspecifically with the membrane, especially under the permeation pressures applied. According to Altmann et al., the deposition of a streaming particle in a crossflow microfluidic system depends on the balance of the lift force and drag force acting on the particle [39]. These hydrodynamic forces are functions of the particle size and flow rate. At low flow rates, all particles have a great chance to interact with the membrane, and nonspecific deposit in this case competes with specific capture of viral particles. On the other hand, high flow rates create great wall shear stress and short sample residence time, both of which lead to reduced binding of both specific and nonspecific species [32]. Thus, the peak capture efficiency at the flow rates 1700–2400 μL/h represents a balance where nonspecific binding is reduced by tangential flow, while specific binding is not yet reduced by the shear stress.

In our tangential flow microfiltration device, membrane clogging is alleviated compared to dead end microfiltration, which often requires vibration or back flow periodically to disperse the cake layer [40–42]. Permeation is well controlled at flow rates up to 3100 μL/h and virus concentrations up to 1×10^7 particles/mL. Combined with affinity capture, close to 100% virus is consistently captured on the membrane from the permeate. Compared to other viral separation devices reported in the literature, the devices described here have the advantages of greater capture yield, high operable flow rates and simple operation procedures [43–45].

Although biotinylated HIV particles and a NeutrAvidin coating are used in this work as proof of principle, the tangential filtration device is adaptable to other affinity chemistries targeting viral surface proteins. As permeation is easily controlled by the geometry of the resistance channel, the capture yield from the input stream is tunable and can be improved by either increasing the permeate fraction or feeding the retentate into the device repeatedly. As the target viral particles are captured through an affinity chemistry on a membrane within a microfluidic channel, enrichment and purification are both achieved. With recent development of optical and electrical bio-sensors [46,47], the filtration device could be developed into a diagnostic device by integrating sensors onto the capture membrane, an ongoing effort in the group. With simplified operation and fast sample processing speed, the device holds promise for viral analysis in resource-limited settings.

5. Conclusions

In conclusion, we successfully assembled a sandwich PDMS device incorporating a nanoporous PC membrane for viral particle separation. Combining tangential filtration and affinity capture, separation of HIV virus is demonstrated in continuous flow. The efficiency of viral capture is flow rate-dependent. At the optimal flow rate, close to 100% of viral particles are captured on the filtration membrane from the permeate with various viral concentrations. This low-cost and easy-to-operate device provides a promising solution to viral sample processing in resource-limited settings.

Author Contributions: All authors contributed to designing the experiments and simulations. Y.W. and K.K. carried out the experiments and analyzed the data. Y.W. performed the simulations. X.C. supervised the project. The original draft was written by Y.W. and all authors contributed to the final manuscript.

Funding: This research was funded by the National Science Foundation, award number CBET-1511284 and National Institutes of Health, grant number NIAID-1R21AI081638.

Acknowledgments: The authors would like to thank Guanyang Xue for photos of the microfluidic device. The authors also acknowledge help with experimental testing from Kieren Connor, Qiongyu Guo, Emily Krulik and Caroline Cai.

Conflicts of Interest: The authors declare no conflict of interest.

References

1. Chin, C.D.; Linder, V.; Sia, S.K. Lab-on-a-chip devices for global health: Past studies and future opportunities. *Lab Chip* **2007**, *7*, 41–57. [CrossRef] [PubMed]
2. Chin, C.D.; Laksanasopin, T.; Cheung, Y.K.; Steinmiller, D.; Linder, V.; Parsa, H.; Wang, J.; Moore, H.; Rouse, R.; Umviligihozo, G.; et al. Microfluidics-based diagnostics of infectious diseases in the developing world. *Nat. Med.* **2011**, *17*, 1015–1138. [CrossRef] [PubMed]
3. Damhorst, G.L.; Watkins, N.N.; Bashir, R. Micro- and Nanotechnology for HIV/AIDS Diagnostics in Resource-Limited Settings. *IEEE Trans. Biomed. Eng.* **2013**, *60*, 715–726. [CrossRef] [PubMed]
4. Ribeiro, R.M.; Qin, L.; Chavez, L.L.; Li, D.F.; Self, S.G.; Perelson, A.S. Estimation of the Initial Viral Growth Rate and Basic Reproductive Number during Acute HIV-1 Infection. *J. Virol.* **2010**, *84*, 6096–6102. [CrossRef]
5. Arraud, N.; Linares, R.; Tan, S.; Gounou, C.; Pasquet, J.M.; Mornet, S.; Brisson, A.R. Extracellular vesicles from blood plasma: determination of their morphology, size, phenotype and concentration. *J. Thromb. Haemost* **2014**, *12*, 614–627. [CrossRef] [PubMed]
6. Sajeesh, P.; Sen, A.K. Particle separation and sorting in microfluidic devices: A review. *Microfluid Nanofluid* **2014**, *17*, 1–52. [CrossRef]
7. Andersson, H.; van den Berg, A. Microfluidic devices for cellomics: A review. *Sens. Actuators B Chem.* **2003**, *92*, 315–325. [CrossRef]
8. Pamme, N. Continuous flow separations in microfluidic devices. *Lab Chip* **2007**, *7*, 1644–1659. [CrossRef] [PubMed]
9. Martinez, A.W.; Phillips, S.T.; Whitesides, G.M.; Carrilho, E. Diagnostics for the Developing World: Microfluidic Paper-Based Analytical Devices. *Anal. Chem.* **2010**, *82*, 3–10. [CrossRef]
10. Magro, L.; Escadafal, C.; Garneret, P.; Jacquelin, B.; Kwasiborski, A.; Manuguerra, J.C.; Monti, F.; Sakuntabhai, A.; Vanhomwegen, J.; Lafaye, P.; et al. Paper microfluidics for nucleic acid amplification testing (NAAT) of infectious diseases. *Lab Chip* **2017**, *17*, 2347–2371. [CrossRef]
11. Seok, Y.; Joung, H.A.; Byun, J.Y.; Jeon, H.S.; Shin, S.J.; Kim, S.; Shin, Y.B.; Han, H.S.; Kim, M.G. A Paper-Based Device for Performing Loop-Mediated Isothermal Amplification with Real-Time Simultaneous Detection of Multiple DNA Targets. *Theranostics* **2017**, *7*, 2220–2230. [CrossRef] [PubMed]
12. Tang, R.H.; Yang, H.; Choi, J.R.; Gong, Y.; Hu, J.; Wen, T.; Li, X.J.; Xu, B.; Mei, Q.B.; Xu, F. Paper-based device with on-chip reagent storage for rapid extraction of DNA from biological samples. *Microchim Acta* **2017**, *184*, 2141–2150. [CrossRef]

13. McDonald, J.C.; Duffy, D.C.; Anderson, J.R.; Chiu, D.T.; Wu, H.; Schueller, O.J.; Whitesides, G.M. Fabrication of microfluidic systems in poly(dimethylsiloxane). *Electrophoresis* **2000**, *21*, 27–40. [CrossRef]

14. Sia, S.K.; Whitesides, G.M. Microfluidic devices fabricated in poly(dimethylsiloxane) for biological studies. *Electrophoresis* **2003**, *24*, 3563–3576. [CrossRef] [PubMed]

15. Kumar, S.; Kumar, S.; Ali, M.A.; Anand, P.; Agrawal, V.V.; John, R.; Maji, S.; Malhotra, B.D. Microfluidic-integrated biosensors: Prospects for point-of-care diagnostics. *Biotechnol. J.* **2013**, *8*, 1267–1279. [CrossRef]

16. Xi, H.D.; Zheng, H.; Guo, W.; Ganan-Calvo, A.M.; Ai, Y.; Tsao, C.W.; Zhou, J.; Li, W.H.; Huang, Y.Y.; Nguyen, N.T.; et al. Active droplet sorting in microfluidics: A review. *Lab Chip* **2017**, *17*, 751–771. [CrossRef]

17. McGrath, J.; Jimenez, M.; Bridle, H. Deterministic lateral displacement for particle separation: A review. *Lab Chip* **2014**, *14*, 4139–4158. [CrossRef]

18. Lenshof, A.; Laurell, T. Continuous separation of cells and particles in microfluidic systems. *Chem. Soc. Rev.* **2010**, *39*, 1203–1217. [CrossRef]

19. Gijs, M.A.; Lacharme, F.; Lehmann, U. Microfluidic applications of magnetic particles for biological analysis and catalysis. *Chem. Rev.* **2010**, *110*, 1518–1563. [CrossRef]

20. Lien, K.Y.; Lin, J.L.; Liu, C.Y.; Lei, H.Y.; Lee, G.B. Purification and enrichment of virus samples utilizing magnetic beads on a microfluidic system. *Lab Chip* **2007**, *7*, 868–875. [CrossRef]

21. Chen, G.D.; Alberts, C.J.; Rodriguez, W.; Toner, M. Concentration and purification of human immunodeficiency virus type 1 virions by microfluidic separation of superparamagnetic nanoparticles. *Anal. Chem.* **2010**, *82*, 723–728. [CrossRef]

22. Surawathanawises, K.; Wiedorn, V.; Cheng, X.H. Micropatterned macroporous structures in microfluidic devices for viral separation from whole blood. *Analyst* **2017**, *142*, 2220–2228. [CrossRef]

23. Chen, G.D.; Fachin, F.; Fernandez-Suarez, M.; Wardle, B.L.; Toner, M. Nanoporous Elements in Microfluidics for Multiscale Manipulation of Bioparticles. *Small* **2011**, *7*, 1061–1067. [CrossRef] [PubMed]

24. Burnouf, T. Modern plasma fractionation. *Transfus. Med. Rev.* **2007**, *21*, 101–117. [CrossRef]

25. Burnouf, T.; Radosevich, M. Nanofiltration of plasma-derived biopharmaceutical products. *Haemophilia* **2003**, *9*, 24–37. [CrossRef]

26. VanDelinder, V.; Groisman, A. Separation of plasma from whole human blood in a continuous cross-flow in a molded microfluidic device. *Anal. Chem.* **2006**, *78*, 3765–3771. [CrossRef]

27. VanDelinder, V.; Groisman, A. Perfusion in microfluidic cross-flow: Separation of white blood cells from whole blood and exchange of medium in a continuous flow. *Anal. Chem.* **2007**, *79*, 2023–2030. [CrossRef]

28. Chen, X.; Cui, D.F.; Liu, C.C.; Li, H. Microfluidic chip for blood cell separation and collection based on crossflow filtration. *Sens. Actuators B Chem.* **2008**, *130*, 216–221. [CrossRef]

29. Wickramasinghe, S.R.; Kalbfuss, B.; Zimmermann, A.; Thom, V.; Reichl, U. Tangential flow microfiltration and ultrafiltration for human influenza A virus concentration and purification. *Biotechnol. Bioeng.* **2005**, *92*, 199–208. [CrossRef]

30. Li, X.; Chen, W.Q.; Liu, G.Y.; Lu, W.; Fu, J.P. Continuous-flow microfluidic blood cell sorting for unprocessed whole blood using surfacemicromachined microfiltration membranes. *Lab Chip* **2014**, *14*, 2565–2575. [CrossRef]

31. Aran, K.; Fok, A.; Sasso, L.A.; Kamdar, N.; Guan, Y.L.; Sun, Q.; Undar, A.; Zahn, J.D. Microfiltration platform for continuous blood plasma protein extraction from whole blood during cardiac surgery. *Lab Chip* **2011**, *11*, 2858–2868. [CrossRef]

32. Mittal, S.; Wong, I.Y.; Deen, W.M.; Toner, M. Antibody-functionalized fluid-permeable surfaces for rolling cell capture at high flow rates. *Biophysical. J.* **2012**, *102*, 721–730. [CrossRef] [PubMed]

33. Chueh, B.H.; Huh, D.; Kyrtsos, C.R.; Houssin, T.; Futai, N.; Takayama, S. Leakage-free bonding of porous membranes into layered microfluidic array systems. *Anal. Chem.* **2007**, *79*, 3504–3508. [CrossRef] [PubMed]

34. Hu, Y.; Cheng, X.H.; Ou-Yang, H.D. Enumerating virus-like particles in an optically concentrated suspension by fluorescence correlation spectroscopy. *Biomed. Opt. Express* **2013**, *4*, 1646–1653. [CrossRef] [PubMed]

35. Chan, L.; Nesbeth, D.; MacKey, T.; Galea-Lauri, J.; Gaken, J.; Martin, F.; Collins, M.; Mufti, G.; Farzaneh, F.; Darling, D. Conjugation of lentivirus to paramagnetic particles via nonviral proteins allows efficient concentration and infection of primary acute myeloid leukemia cells. *J. Virol.* **2005**, *79*, 13190–13194. [CrossRef] [PubMed]

36. Amiri, A.; Vafai, K. Transient analysis of incompressible flow through a packed bed. *Int. J. Heat Mass Tran.* **1998**, *41*, 4259–4279. [CrossRef]

37. Zhang, W.Y.; Ferguson, G.S.; Tatic-Lucic, S. Elastomer-supported cold welding for room temperature wafer-level bonding. In Proceedings of the 17th IEEE International Conference on Micro Electro Mechanical Systems, Maastricht, The Netherlands, 25–29 January 2004; pp. 741–744.

38. Cheng, X.H.; Irimia, D.; Dixon, M.; Sekine, K.; Demirci, U.; Zamir, L.; Tompkins, R.G.; Rodriguez, W.; Toner, M. A microfluidic device for practical label-free CD4+T cell counting of HIV-infected subjects. *Lab Chip* **2007**, *7*, 170–178. [CrossRef] [PubMed]

39. Altmann, J.; Ripperger, S. Particle deposition and layer formation at the crossflow microfiltration. *J. Membrane Sci.* **1997**, *124*, 119–128. [CrossRef]

40. Parameshwaran, K.; Fane, A.G.; Cho, B.D.; Kim, K.J. Analysis of microfiltration performance with constant flux processing of secondary effluent. *Water Res.* **2001**, *35*, 4349–4358. [CrossRef]

41. Yoon, Y.; Kim, S.; Lee, J.; Choi, J.; Kim, R.K.; Lee, S.J.; Sul, O.; Lee, S.B. Clogging-free microfluidics for continuous size-based separation of microparticles. *Sci. Rep.* **2016**, *6*, 26531. [CrossRef]

42. Chellam, S.; Jacangelo, J.G.; Bonacquisti, T.P. Modeling and experimental verification of pilot-scale hollow fiber, direct flow microfiltration with periodic backwashing. *Environ. Sci. Technol.* **1998**, *32*, 75–81. [CrossRef]

43. Kim, Y.G.; Moon, S.; Kuritzkes, D.R.; Demirci, U. Quantum dot-based HIV capture and imaging in a microfluidic channel. *Biosens. Bioelectron.* **2009**, *25*, 253–258. [CrossRef] [PubMed]

44. Furtak, V.A.; Dabrazhynetskaya, A.; Volokhov, D.V.; Chizhikov, V. Use of tangential flow filtration for improving detection of viral adventitious agents in cell substrates. *Biologicals* **2015**, *43*, 23–30. [CrossRef] [PubMed]

45. Wang, S.; Sarenac, D.; Chen, M.H.; Huang, S.H.; Giguel, F.F.; Kuritzkes, D.R.; Demirci, U. Simple filter microchip for rapid separation of plasma and viruses from whole blood. *Int. J. Nanomedicine* **2012**, *7*, 5019–5028. [CrossRef] [PubMed]

46. Yoo, S.M.; Lee, S.Y. Optical biosensors for the detection of pathogenic microorganisms. *Trends Biotechnol.* **2016**, *34*, 7–25. [CrossRef] [PubMed]

47. Tokel, O.; Yildiz, U.H.; Inci, F.; Durmus, N.G.; Ekiz, O.O.; Turker, B.; Cetin, C.; Rao, S.; Sridhar, K.; Natarajan, N.; et al. Portable microfluidic integrated plasmonic platform for pathogen detection. *Sci. Rep.* **2015**, *5*, 9152. [CrossRef] [PubMed]

 *micromachines*

Article

Portable Rice Disease Spores Capture and Detection Method Using Diffraction Fingerprints on Microfluidic Chip

Ning Yang [1], Chiyuan Chen [1], Tao Li [1], Zhuo Li [1], Lirong Zou [1], Rongbiao Zhang [1,*] and Hanping Mao [2,*]

[1]   School of Electrical and Information Engineering, Jiangsu University, Zhenjiang 212013, China; yangning7410@163.com (N.Y.); chenchiyuan0214@163.com (C.C.); litao0203a@163.com (T.L.); 13037591303@163.com (Z.L.); wxzlr1996@163.com (L.Z.)

[2]   School of Agricultural Equipment Engineering, Jiangsu University, Zhenjiang 212013, China

*   Correspondence: zrb@ujs.edu.cn (R.Z.); hanpingmao1811@163.com (H.M.)

Received: 3 April 2019; Accepted: 25 April 2019; Published: 28 April 2019

Abstract: Crop diseases cause great harm to food security, 90% of these are caused by fungal spores. This paper proposes a crop diseases spore detection method, based on the lensfree diffraction fingerprint and microfluidic chip. The spore diffraction images are obtained by a designed large field of view lensless diffraction detection platform which contains the spore enrichment microfluidic chip and lensless imaging module. By using the microfluidic chip to enrich and isolate spores in advance, the required particles can be captured in the chip enrichment area, and other impurities can be filtered to reduce the interference of impurities on spore detection. The light source emits partially coherent light and irradiates the target to generate diffraction fingerprints, which can be used to distinguish spores and impurities. According to the theoretical analysis, two parameters, Peak to Center ratio (PCR) and Peak to Valley ratio (PVR), are found to quantify these spores. The correlation coefficient between the detection results of rice blast spores by the constructed device and the results of microscopic artificial identification was up to 0.99, and the average error rate of the proposed device was only 5.91%. The size of the device is only 4 cm × 4 cm × 5 cm, and the cost is less than $150, which is one thousandth of the existing equipment. Therefore, it may be widely used as an early detection method for crop disease caused by spores.

Keywords: crop disease; lensfree; light diffraction; image processing; microfluidic

---

1. Introduction

Rice is the most important crop all around the world. Currently, approximately 5.6 billion people (80% of the world's population) have rice as their staple food. Rice consumption is expected to reach 8 million tons in 2030 owing to population and economic growth [1]. The blast fungus causes a serious disease on a wide variety of grasses including rice, wheat, and barley [2]. Rice blast disease, which is caused by a kind of spore named Magnaportheoryzae [3], affects most of the rice-producing countries and has already spread to approximately 85 countries [4]. The severity of rice diseases has been unprecedented in the past 20 years, particularly in China's Yangtze River basin region in 2018. Most rice disease spores are airborne microparticles with a size of 10–20 μm, and are difficult to detect and capture. Therefore, it is necessary to establish a rapid and effective spore detection facility to prevent crop diseases.

Several methods have been proposed to detect spores which cause rice diseases. Spore analysis is a powerful technique and has been widely used in plant diseases control [5]. In these researches, identification of spores, which can quantitatively evaluate the effect of disease control, is a routine and essential tache in spore assays [6]. Polymerase chain reaction (PCR) is widely used in spore detection due to its high quantitative analysis ability [7]. Rice is a widely cultivated food crop. Although the accuracy of PCR detection is high and the speed is fast enough, the method needs professional personnel and to be sent back to the laboratory for testing, which is not suitable for large-scale promotion in the field. The staining method is also often used in cell and spore identification. The staining agent can impregnate the target and mark it [8], which requires various special stains and destroys the tissue smears. Micromechanical cantilever arrays can be used to detect spore concentration rapidly and quantitatively. This method has the characteristics of high precision and high sensitivity, but it requires a clean detection environment [9]. The common method is to visually measure and count the spores under microscope [10], but the microscope is expensive and bulky, and cannot be widely used outdoors. Most of these methods require complex procedures and multiple reagents, resulting in low experimental repeatability [11]. Also, these complex experimental procedures are time-wasting and labor-intensive [12].

To improve the above problems, developing a convenient and intelligent low-cost spore detection facility has become a hotspot of the filed. Lensless imaging technology is valuable for spore research because in these research fields the statistical regularities are based on the behavior of large populations. Therefore, it is concerned by researchers and applied in the fields of blood cell counting [13] and viability detection [14], particle size classification [15], biofilm detection [16], etc. Traditional optical microscopy has a complex lens structure, so it is not suitable for large-scale popularization because of its large volume and high price. Its precise optical part is sensitive to external conditions such as temperature and humidity, and it is difficult to adapt to complex field environments. Compared with the microscopic imaging mode, lensless imaging technology only uses a complementary metal oxide semiconductor (CMOS) module, which is small in size, stable in performance, and not susceptible to environmental impact. Traditional microscopic imaging has a very small field of view (FOV), usually less than 0.5 mm^2. Without moving the lens, it is difficult to cover the whole enrichment area of microfluidic chip. The active imaging area of the CMOS sensor is about 20 mm^2, which means the FOV is about 20 mm^2 and it is approximately 100 times larger than that of a conventional optical microscope. Different to the field of biomedicine, there are many impurities in the air. If the spores are collected directly without filtering, a large number of impurities will be mixed into the samples, which will reduce the accuracy of image detection in the later stage. Therefore, the microfluidic chip must be added to the spore collection process to separate most impurities. Microfluidic chip has many advantages, such as little reagent consumption, fast automatic detection, low cost, high integration, and has allowed researchers concentrate the whole process of a biochemical reaction on the chip [17]—mostly applied in the field of liquid and medical testing [18]. Because PM2.5, PM10, and other air pollutants have a great impact on human health, researchers have designed a variety of microfluidic chips for the filtration and enrichment of airborne microparticles [19,20]. However, few microfluidic chips have been developed to enrich spores of crop diseases. In this paper, a microfluidic chip is designed to enrich and separate airborne spores. According to the above technology, a spore detection device has been developed, which includes the microfluidic chip and lensless imaging module. It can enrich and detect spores and realize early monitoring of crop diseases.

2. Materials and Methods

2.1. Design of Microfluidic Chip

The microfluidic device is fabricated by the standard soft-lithography technique [21,22]. The master mold is created on a silicon substrate mold by exposing the photoresist SU-8. After being peeled off from the mold, the polydimethylsiloxane (PDMS) slab is punched through to make ports at the inlet and outlet. The plastic tubes, with a small amount of glue at their ends, are inserted through the inlet and outlet ports. The PDMS slab is treated with oxygen plasma and then bonded to a glass substrate (20 mm × 55 mm). Finally, the assembled device is cured at 70 °C for 30 min to enhance bonding. A sectional view of the enrichment and separation of the microfluidic chip is shown in Figure 1. It has a two-stage separation–enrichment structure to simplify the fabrication process and reduce the costs. L1, L2, L3, and L4 represent the width of channel 2, channel 3, channel 4, and channel 5, respectively. The value of D1 is collection tank 1 entrance width minus channel 2 width. The value of D2 is width of collection tank 2 entrance minus width of channel 4. The arc channel is composed of two concentric arcs with different radii. R1 and R2 are the radius of the inner arc of the arc channels respectively. To obtain better experimental results, 16 µm particles were used to represent rice blast spores and 2 µm particles were utilized to represent impurities in the environment. When particles pass through channel 1 and Arc channel 1, the force on the particles is the same because of the same flow rate of the two channels, but the direction of the force was changed due to the change of the shape. The two particles with different sizes deviate from the trajectory due to the force in two different directions, thus realizing the original separation of the two particles. The 16 µm particles with a larger particle size enter collection tank 1 due to their larger velocity. In order to keep the 16 µm particles in collection tank 1 and 2 µm particles into the second structure, L1 should be smaller than L2. The shape and design principle of the first and second structure are the same, so L3 should be smaller than L4.

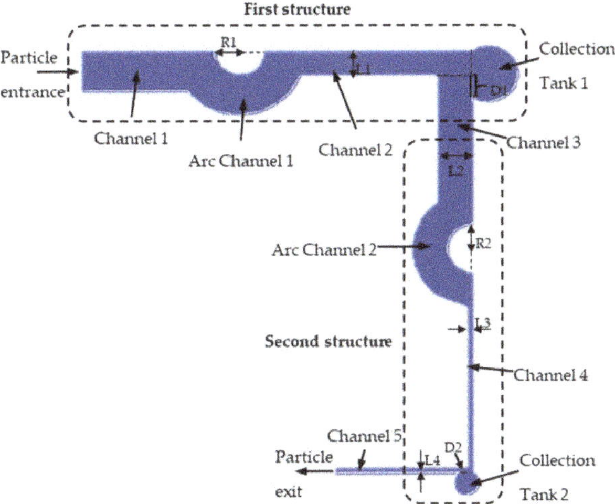

Figure 1. Structural chart of microfluidic chip.

2.2. Numerical Method

The commercial COMSOL Multiphysics 5.1 was utilized to simulate the flow field and particle motion of microfluidic chip. For numerical analysis, the model parameters were set at width of particle entrance = 2000 µm, length of channel 1 = 12,000 µm, R1 = 1000 µm, width of Arc channel 1 = 2000 µm, length of channel 3 = 8000 µm, R2 = 1050 µm, width of Arc channel 2 = L2, length of channel 4 = 10,000 µm, length of channel 5 = 5000 µm. All microchannels are 100 µm in height. The parametric study was performed by varying the L2/L1, L4/L3 ratio in the range of 1–2 and D1, D2 value in the range of 800–1800 µm to determine the effect of enrichment and separation.

The flow was assumed to be steady, two-dimensional axisymmetric, and incompressible. Moreover, Reynolds numbers of collection tank 1 and 2 were 173 and 346, respectively. The temperature was 293.15 K and the applied pressure was 101.3 kPa. The velocity inlet condition at the particle entrance, pressure outlet condition at the particle exit, and no-slip condition in all walls were set as the boundary conditions. The laminar inflow rate is 115 mL/min. Continuity, momentum, and energy equations were solved repeatedly by adjusting the convergence criterion at 10^{-6}.

After flow analysis, the behavior of particles was investigated. For the particle behavior analysis, the COMSOL Multiphysics Particle Tracing Module was used. A particle was assumed to be permanently collected when it hit any wall. The particle density was set at 1000 kg/m^3 to express the aerodynamic size. Spore concentration in the air is very low, so 40 particles (2 μm particles and 16 μm particles, 20 of each) were arranged at regular intervals from the center of the particle entrance to the edge. The Lagrangian method is convenient for describing particle movement. The single particle motion equation is expressed as follows:

$$m_p \frac{d^2 r_p}{dt^2} = F_f + F_o.$$

$$(1)$$

In the formula, r_p is the particle motion position vector, m_p is the particle mass, and t is the particle motion time. F_f is the force of fluid on particles and F_o is the force of an external potential field on particles. The particle motion can be followed by solving the particle motion Equation (1).

2.3. Diffraction Imaging Detection Platform Setup

The detection facility designed in this paper has the characteristics of small size, low-cost, and convenience. The schematic diagram is shown in Figure 2. The facility has main two parts—the microfluidic chip and the lensless imaging module. The nano-microspheres (1–4 μm), pollen, dust, and spore-carrying rice grains were mixed with water and put into the aerosol generator to produce aerosol. Then, the aerosol was passed through the drying tube to eliminate the water vapor. Then, the dried mixture gas was passed through the flowmeter to obtain stable gas flow rate. At this stage the mixed gas finally enters the microfluidic chip.

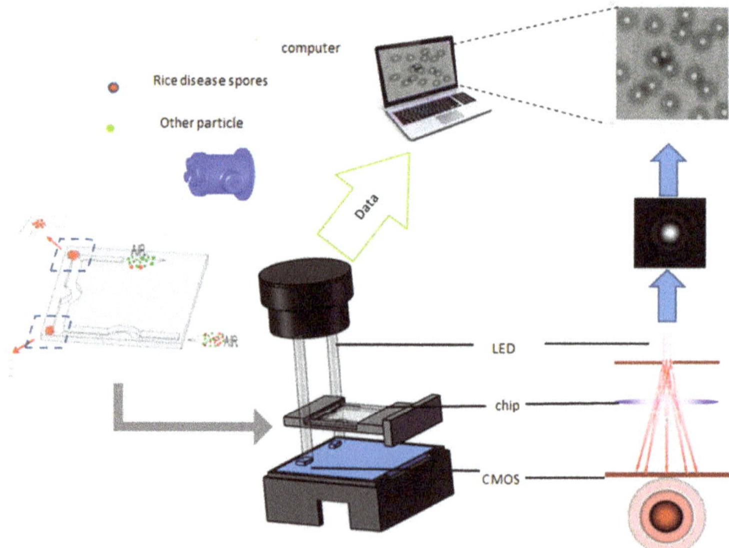

Figure 2. Portable real-time rice disease spores detection using characteristics of diffraction fingerprints on the microfluidic chip.

The spores were separated and enriched by external force through the microfluidic chip, shown in Figure 2. The microfluidic chip consists of two red-labeled enrichment regions and three microchannels with different diameters. The lensless imaging module includes the light source and CMOS sensor. OSRAM's LA E65B LED (617 nm) is selected as the light source. The diameter of the micropore was

100 μm directly below it. The CMOS sensor was located 45 mm below the micropore. The MT9P031 image sensor chip with 5 million pixels from Aptina was selected. The size of the imaging area was 5.70 mm × 4.28 mm, which means the FOV was up to 24 mm^2 and approximately 100 times larger than that of a 200× conventional microscope. The LED light source was monochrome, so a black-and-white version of the chip was customized to obtain gray image directly to improve the signal-to-noise ratio of the image. In addition, a drawer-type sample tray was designed to realize quick replacement of the microfluidic chip. When using the device, the air pump was turned on to collect the spore mixture gas into the microfluidic chip for enrichment and separation. Then, the lensless imaging module photographed the diffraction fingerprints of enriched spores in the microfluidic chip and transferred the image to PC for image processing and saving. The facility realizes the independent operation of spore enrichment, sampling, photographing, uploading, and detection.

2.4. Spore Detection Using Diffraction Parameters

In order to design a spore detection system based on spore diffraction fingerprint, the acquisition and image processing of the diffraction fingerprint were automatic. The steps of acquisition and analysis of lensless diffraction fingerprint information are as follows: (1) Spores are enriched by microfluidic chip; (2) the enrichment area of microfluidic chip is placed on the CMOS sensor; (3) high definition diffraction fingerprint images are acquired by CMOS sensor; (4) diffraction fingerprint images are transferred to computer for further processing as described in the following steps: a. Images are read sequentially; b. the diffraction fingerprints of each spore are extracted from the picture; c. characteristic parameters are calculated by using pixel information of diffraction fingerprint; d. comparing the calculated characteristic parameters with the threshold value of spore characterization to determine whether the target is the spore or not. There is no need to add additional reagents and the whole process is time-saving and labor-saving.

In the study (shown in Figure 3), two parameters, Peak to Center ratio (PCR) and Peak to Valley ratio (PVR), were designed to quantify these spore fingerprint characteristics and are formulized as follows, respectively:

$$PCR = \frac{AP}{C}, \tag{2}$$

$$PVR = \frac{AP}{AV}. \tag{3}$$

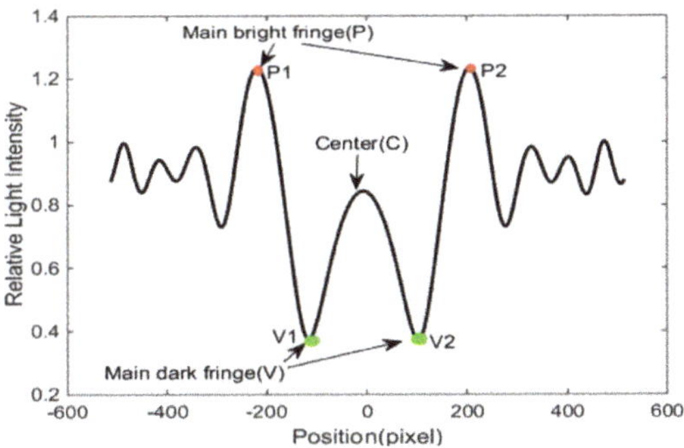

Figure 3. The light intensity value profile of calculated spore diffraction fingerprint.

AP is the average relative light intensity of the two peaks (P1, P2) in a single spore fraction fingerprint. AV is the average relative light intensity of the two lowest points (V1, V2). The dark fringe nearest to the center (C) is called the main dark fringe, and the bright fringe is called the main bright fringe.

The parameters were employed to distinguish spores and other impurities. Matlab was utilized to extract spore fingerprints and calculate the parameters corresponding to each spore. A spore was identified as a rice blast spore only when the two calculated parameters satisfy both of the two conditions that PCR and PVR were between the determined threshold. The threshold levels of every specific spore lines should be determined by comparing the average PCRs and PVRs of rice blast spores in advance. In the study, the two threshold levels of the selected rice blast spores' lines were determined to be 1.25–1.40 of PCR and 3.35–3.5 of PVR from 2000 spores of rice blast. The spore, whose corresponding fingerprint parameters satisfy the thresholds given above, could be identified as a rice blast spore.

3. Results and Discussion

3.1. Simulation and Experiment of Microfluidic Chip

3.1.1. Particle Motion Simulation

Figure 4a,b shows the relationship between the separation and enrichment effect of 16 μm particles and L1, and the relationship between the separation and enrichment effect of 16 μm particles and L2/L1. The D1 is 500 μm, the L1 and value of L2/L1 is changed. The separation and enrichment effect of 16 μm particles are better when L1 = 1500, 1400, and 1300 μm; and the separation and enrichment effect of 16 μm particles are best when L2/L1 = 1.4.

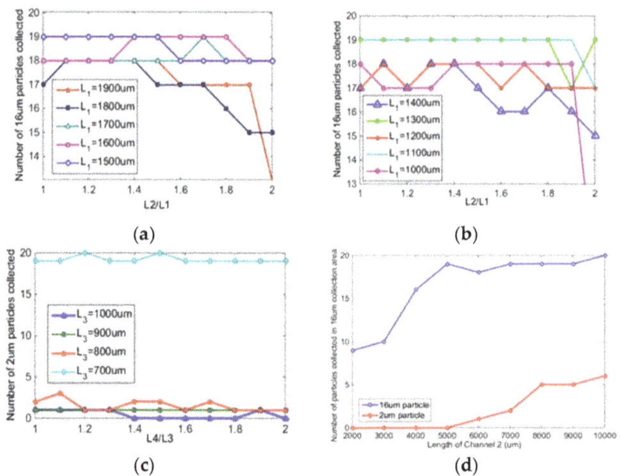

Figure 4. (a,b) Separation and enrichment of 16 μm particles; (c) separation and enrichment of 2 μm particles; (d) effect of channel 2 length on particle separation and enrichment.

In order to study the relationship between the separation and enrichment effect of 2 μm particles and L3, the relationship between the separation and enrichment effect of 2 μm particles and L4/L3 must also be studied. Figure 4c only shows L3 = 700, 800, 900, 1000 μm because when L3 takes other larger values, 2 μm particles cannot be collected, but all reach the exit. When L3 = 700 μm, the separation and enrichment effect of 2 μm particles are better regardless of the value of L4/L3; when L3 takes different values, there is no uniform L4/L3 value that can make the best separation and enrichment effect under these conditions.

In order to study the effect of channel 2 length on particle enrichment and separation, a variable channel 2 length (2000–10,000 μm) was set up. As shown in Figure 4d, when the length of channel 2 is 5000 μm, 6000 μm, and 7000 μm, most 16 μm particles can enter the 16 μm collection tank, and most

2 μm particles can enter the next structure. When channel 2 is too short, all particles cannot obtain enough centrifugal force, so it is difficult for particles to enter the collection area. When channel 2 is too long, the centrifugal force obtained by the particles will be too large, and the small particles will also enter the 16 μm collection area. Finally, the length of channel 2 is 5000 μm.

In order to study the relationship between the separation and enrichment effect of 16 μm particles and the D1, the L2/L1 is 1.4. By changing the value of D1, the range of D1 is 100–1000 μm. When L1 changes from 1900 μm to 1000 μm, the number of 16 μm particles collected is studied. Figure 5a,b shows that the D1 of optimum separation and enrichment is different when L1 is different, and 200, 300, and 500 μm can be regarded as the optimum D1.

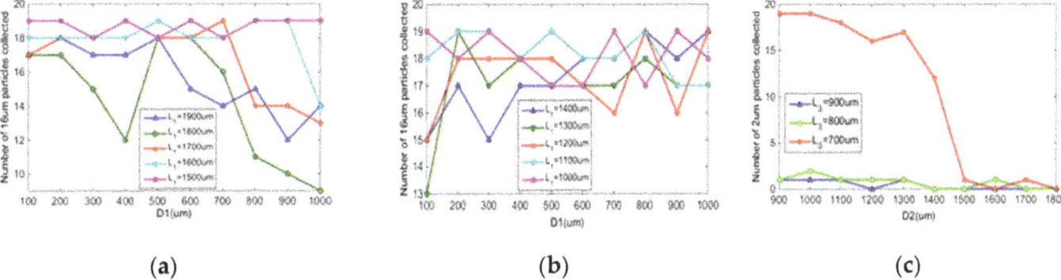

(a)   (b)   (c)

Figure 5. (**a**,**b**) Relationship between the separation and enrichment effect of 16 μm particles and D1; (**c**) relationship between the separation and enrichment effect of 2 μm particles and D2.

Figure 5c shows the relationship between the separation and enrichment effect of 2 μm particles and D2. The number of 2 μm particles collected under different L3 conditions was studied. The best separation and enrichment effect are obtained at D2 = 1000 μm.

In order to determine the final structure of microfluidic chip, the optimal structure parameters should be further selected. For the primary structure of collecting 16 μm particles, the optimum L2/L1 is 1.4. Considering that the small size of micro-structure will increase the complexity of the fabrication process, L1 = 1500 μm is chosen as the optimum channel width and channel 2 is 5000 μm in length. When L1 = 1500 μm, D1 has little effect on the separation and enrichment of particles. In order to simplify the structure, the D1 is 500 μm. For the secondary structure of collecting 2 μm particles, only when L3 = 700 μm can the separation and enrichment effect be better. When L3 = 700 μm, L4/L3 has little effect on the separation and enrichment of particles. In order to simplify the structure, L4/L3 = 1.4 is chosen. The optimum D2 is 1000 μm. Figure 6a is the simulation result of the separation and enrichment effect of two kinds of particles, 20 particles each. 19 particles of 16 μm were collected in collection tank 1, and 19 particles of 2 μm was collected in collection tank 2. Figure 6b is the velocity distribution in the microchannel of the chip.

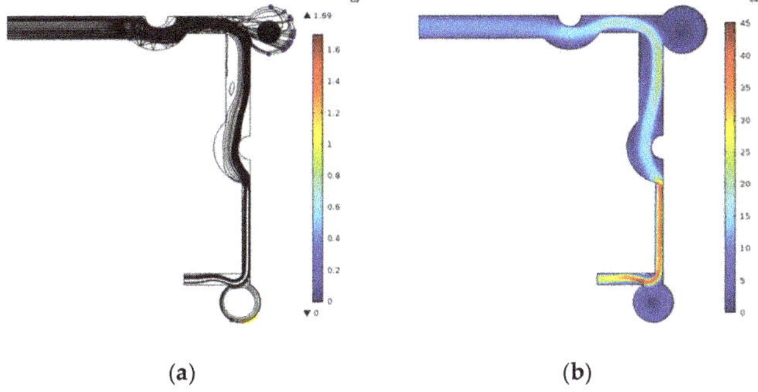

(a)   (b)

Figure 6. (**a**) Particle trajectory simulation in microfluidic chip; (**b**) simulation of velocity distribution in microfluidic chip (m/s).

As shown in Figure 6, the principle of separation and enrichment of the first structure is first analyzed. The mixed gas enters the microchannel from the inlet of the microfluidic chip and obtains a horizontal to right initial velocity. If there is no Arc channel 1, the particles cannot get enough centrifugal force to enter collection tank 1 at a right angle turn. When the Arc channel 1 is added, the Arc channel 1 has a coupling effect with the right angle turn, which increases the centrifugal force of the particles, so that it can enter collection tank 1. In the simulation, the density of all the particles is the same, so the larger the diameter of the particles, the greater the centrifugal force. The 16 μm particles can enter collection tank 1 due to their large centrifugal force, while the 2 μm particles can enter the second structure. Then the principle of separation and enrichment of the second structure is analyzed. Similar to the first structure, Arc channel 2 and the right angle turn produce a coupling effect, which increases the centrifugal force of the particles and facilitates entry into collection tank 2. Reducing L3 accelerates the particles, increasing the acceleration of the 2 μm particles and increasing the centrifugal force to enter collection tank 2.

3.1.2. Spore Collection Experiment

Figure 7a,b is an experimental image taken by a scanning electron microscope. Rice blast spores are collected in collection tank 1. The target in the red circle is the rice blast spore. The spore samples of rice blast were cultured from rice grains, so the fine impurities in collection tank 1 were rice husks, as shown in the black circle marker (not all markers for the clarity of the picture), which affect the shooting effect. Other impurities are collected in collection tank 2. Therefore, the designed microfluidic chip can separate and collect rice blast spores.

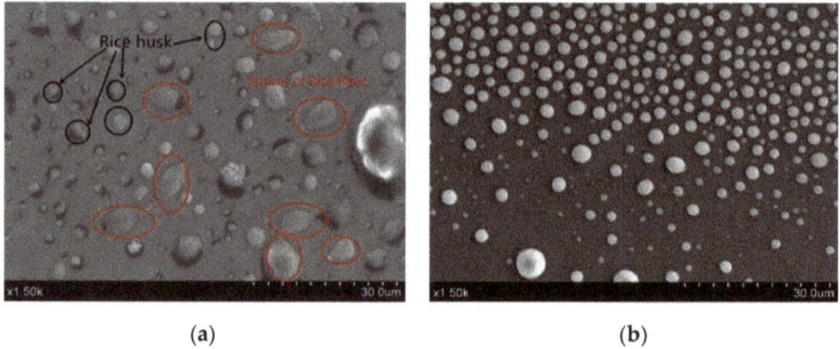

(a) (b)

Figure 7. (a) Samples collected in collection tank 1; (b) samples collected in collection tank 2.

3.2. Diffraction Imaging Detection Platform

3.2.1. Diffraction Fingerprints Calculation and Investigation

Traditional optical microscopy imaging technology is based on the change of wavelength (color) and amplitude (brightness) when light passes through transparent substances. The morphology of microorganisms is directly observed by the naked eye through microscopy. The light emitted by LED is converted into coherent light through micro-holes, thus realizing coherent illumination. As shown in Figure 2, the light emitted by the LED light source passes through the micropore directly below it, and produces partially coherent light. The partially coherent light propagates over a certain distance and then irradiates on the sample plane (microfluidic chip). Diffraction hologram is formed by the interference of the sample's backlight and scattered light on the sample plane, which is photographed by CMOS imaging chip directly below the sample plane.

The classical scalar diffraction theory was first proposed by Huygens in 1678. Huygens believed that any point on the wave surface could be regarded as the source of secondary spherical wavelet in the process of light propagation. Fresnel further proposed in 1818 that the interference between the wavelet source and the wavelet source should be superimposed. Hence, the famous Huygens–Fresnel principle is proposed that the light vibration at any point outside the wavefront should be the result of the coherent superposition of all wavelets on the wavefront. The Huygens–Fresnel principle can be utilized to explain the diffraction of light. Its mathematical expression can be written as follows:

$$U(Q) = c \iint_{\Sigma} U_0(P)K(\theta)\frac{e^{jkr}}{r}dS. \tag{4}$$

Among them, S is the surface of object wavefront. In this work, the object wavefront S is formed by light source transmitting through spore sample. c is a constant, Σ is an integral plane, $U_0(P)$ is the complex amplitude of P at any point on the wavefront, $U(Q)$ is the complex amplitude of Q at any point in the light wave field, r is the distance from P to Q, θ is the angle formed by the normal N and PQ at the wavelet front where P is located, and $K(\theta)$ is the tilt correlation between P and Q.

MATLAB was utilized to calculate the diffraction fingerprint with the same parameters setting as that of the experimental setup. Fourier transform and discretization are used and suitable sampling intervals are selected in calculation. Figure 8a shows that the spore image is processed into an impermeable boundary and analyzed as the pattern of a diffraction screen. Firstly, the contour of the spore is displayed by setting the gray threshold through the grayscale display, then the noise in the image is removed by setting the threshold of the boundary pixel point, and the blank vacancy in the spore is filled. Finally, a similar spore pattern is generated through the black and white inversion. The generated spore contour image is calculated using the formula (4) to obtain the image shown in Figure 8b. The ordinate represents the relative intensity of the light field. It can be seen that the amplitude of the central light field is high, the amplitude of the edge is low, and the periphery of the edge presents a state of high and low alternation. In other words, in the center of the captured pattern the diffraction fringes are distinct, while at the edge the fringes are effectively eliminated. The reason is that the partially coherent light source contains rays of different phase. The rays lead to different fringe distribution at the edge of the pattern. The result is good for diffraction fingerprint quality because it reduces spore's signatures overlap. The center area and these main fringes are more valuable.

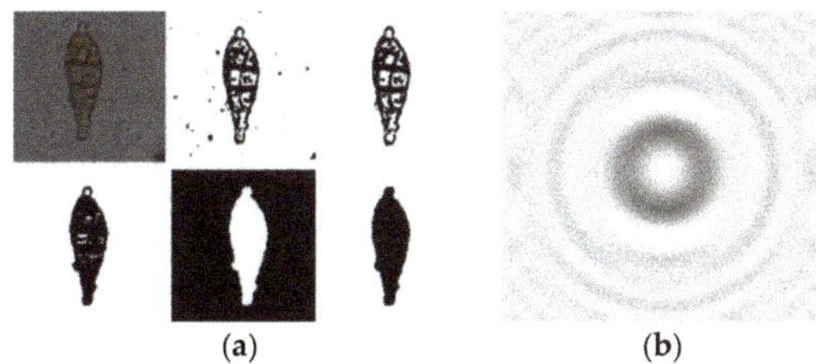

(a)   (b)

Figure 8. Spore image processing and spore fingerprint captured in this work. (**a**) Gray processing of spore image; (**b**) spore diffraction fingerprint calculation.

A rice blast spore can be approximately regarded as an oval. The observations indicate that spore morphology has a close relationship with the diffraction fingerprint. The spore models were set to ellipses, which had the same coverage area but different short and long axes. The ratio of semi-minor axis length is 1:1, 1.5:1, 2:1, 2.5:1, and 3:1. Their diffraction intensity patterns are shown in Figure 9. As the ratio of semi-minor axis length increases, the position of the main dark fringe and the main bright fringe changes little, while the brightness of the main dark fringe of the diffraction fingerprint decreases and the brightness of the main bright fringe increases. The above changes will cause changes in PCR and PVR parameters.

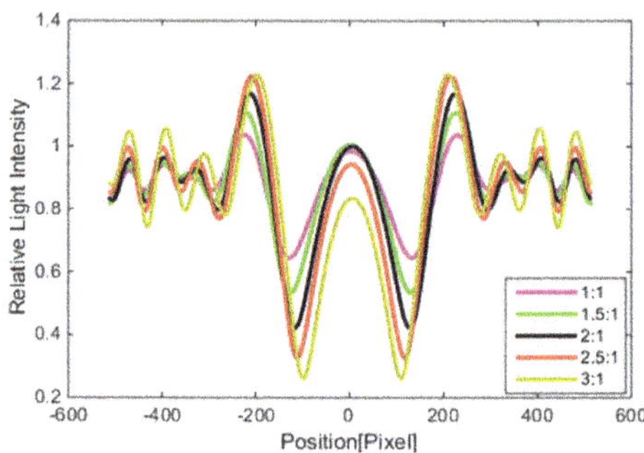

Figure 9. The light intensity value profile of the calculated spore diffraction fingerprint.

According to the analysis above, two parameters, PCR and PVR are designed to quantify the Peak to Center ratio and Peak to Valley ratio. The two computational expressions are described in formulas (1) and (2).

3.2.2. Validation and Superiority of Proposed Approach

Gas mixed rice blast spores were prepared by using an aerosol generator. Then, mixed gas was pumped into the designed microfluidic chip and the performance of the device was verified by taking pictures. Collection tank 2 of the microfluidic chip was observed under a microscope. Figure 10a is a spore sample which has not been enriched and separated by microfluidic chip. There are more small particle impurities in it than in Figure 10b. Excessive impurities will cause adverse effects such as overlap, as Figure 10c shows, which will affect the accuracy of subsequent diffraction imaging detection of spores. The detection accuracy of samples without microfluidic chip enrichment is about 80%, and that of samples with microfluidic chip enrichment is 94%. Because of the high concentration of spores in the disposed spore solution and less other impurities, better results can be obtained without using the microfluidic chip for enrichment and separation. However, in the field environment there are many impurities, low spore concentration and excessive impurities will make the detection accuracy less than 50%. So, it is necessary to use microfluidic chip to pre-filter samples before detection. To assess the validity and reliability of a screening method which used the two parameters as a criterion for spore screening, 20 experimental samples mixed with impurities such as pollen, dust, and beads were selected in the assay. The two parameters, PCR and PVR, were computed from the corresponding diffraction fingerprints. Twenty groups of experiments were conducted. Figure 10d is the result of the test. The manual counting results and automatic counting results are shown in Figure 10e. It can be concluded that the error rate ((Automatic counting result–Manual counting result)/ Manual counting result) fluctuates around 5%, with a maximum of 10% and an average error of 5.91%, which is an acceptable range. The correlation coefficient (R^2) of the 20 groups of test results is 0.9912, and the automatic counting method is highly correlated with the manual counting method. Figure 10f is a cluster analysis of rice blast spores and 2 μm beads using PVR and PCR parameters. It can be concluded that the two substances are clearly classified. Figure 10g is the result of Bland–Altman analysis of the two methods. All the points are within the 95% consistency interval. These analysis results reveal a good agreement between the two modalities and prove the validity of this approach. Furthermore, the proposed approach is a continuous detection method, and the cost of our constructed platform is far lower than that of other techniques.

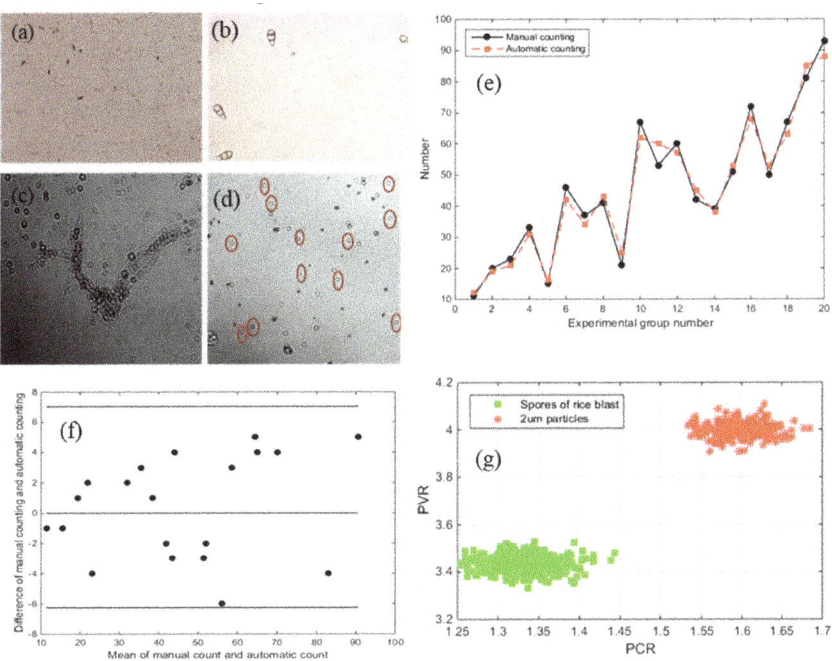

Figure 10. (**a**) Spore samples not enriched and separated by microfluidic chip; (**b**) spore samples enriched and separated by microfluidic chip; (**c**) overlapping interference of diffraction image; (**d**) diffraction image taken by complementary metal oxide semiconductor (CMOS), the fingerprints in red circles correspond to spores; (**e**) experimental result; (**f**) cluster analysis; (**g**) analysis by the Bland–Altman method.

4. Conclusions

In summary, an approach which can enrich and distinguish spores utilizing a microfluidic chip and lensfree spore diffraction fingerprints in an automatic manner is demonstrated. The designed microfluidic chip can enrich rice blast spores. The two parameters, PCR and PVR, were firstly designed to quantify the characteristics of spore diffraction fingerprint and validated to screen spores. Both of the proposed approaches and manual counting methods were employed to measure spore. The high correlation of 0.9912 and Bland–Altman analysis revealed a good agreement and validated the proposed approach. The detection accuracy of rice blast spores was 94%. The cost of our constructed platform is far lower than others, and no extra operation is required during the detection process. This approach is high-automatic, continuous, time-saving, and labor-saving. These results and advantages illustrate that the proposed approach may be widely used for fungal disease detection caused by spores.

Author Contributions: Conceptualization, N.Y.; methodology, C.C.; software, C.C.; validation, T.L. and L.Z.; formal analysis, N.Y.; investigation, L.Z.; resources, N.Y.; data curation, L.Z.; writing—original draft preparation, T.L. and C.C.; writing—review and editing, C.C.; visualization, N.Y.; supervision, N.Y.; project administration, R.Z.; funding acquisition, H.M.

Funding: This research was funded by Chinese National Natural Science Foundation (grant number 31701324); China Postdoctoral Science Foundation Project (grant number 2018M642182); Jiangsu Agricultural Science and Technology Innovation Fund (grant number CX(18)3043); Outstanding Youth Science Foundation of Jiangsu province (grant number BK20180099); Zhenjiang Dantu Science and Technology Innovation Fund (Key R&D Program-Social Development), (grant number SH2018003); the Priority Academic Program Development of Jiangsu Higher Education Institutions (PAPD).

Conflicts of Interest: The authors declare no conflict of interest.

References

1. Wang, G.-L.; Valent, B. Durable resistance to rice blast. *Science* **2017**, *355*, 906–907. [CrossRef] [PubMed]
2. Talbot, N.J. On the trail of a cereal killer: Exploring the biology of Magnaporthe grisea. *Annu. Rev.* **2003**, *57*, 177–202. [CrossRef]
3. Dai, X.; He, C.; Zhou, L.; Liang, M.; Fu, X.; Qin, P.; Yang, Y.; Chen, L. Identification of a specific molecular marker for the rice blast-resistant gene Pigm and molecular breeding of thermo-sensitive genic male sterile leaf-color marker lines. *Mol. Breeding* **2018**, *386*, 72. [CrossRef]
4. Yangseon, K.; Jae-Hwan, R.; Ha, K. Early Forecasting of Rice Blast Disease Using Long Short-Term Memory Recurrent Neural Networks. *Sustainability* **2017**, *10*, 34.
5. Mentlak, T.A.; Kombrink, A.; Shinya, T.; Ryder, L.S.; Otomo, I.; Saitoh, H.; Terauchi, R.; Nishizawa, Y.; Shibuya, N.; Thomma, B.P.H.J.; et al. Effector-Mediated Suppression of Chitin-Triggered Immunity by Magnaporthe oryzae Is Necessary for Rice Blast Disease. *Plant Cell* **2012**, *24*, 322–335. [CrossRef]
6. Li, Y.; Lu, Y.-G.; Shi, Y.; Wu, L.; Xu, Y.-J.; Huang, F.; Guo, X.-Y.; Zhang, Y.; Fan, J.; Zhao, J.-Q.; et al. Multiple Rice MicroRNAs Are Involved in Immunity against the Blast Fungus Magnaporthe oryzae. *Plant Physiol.* **2014**, *164*, 1077–1092. [CrossRef]
7. Chen, J.J.; Li, K.T. Analysis of PCR Kinetics inside a Microfluidic DNA Amplification System. *Micromachines* **2018**, *9*, 48. [CrossRef]
8. Chesmore, D.; Bernard, T.; Inman, A.J.; Bowyer, R.J. Image analysis for the identification of the quarantine pest Tilletia indica*. *EPPO Bull.* **2003**, *33*, 495–499. [CrossRef]
9. Osimani, A.; Milanovi, V.; Cardinali, F.; Garofalo, C.; Clementi, F.; Pasquini, M.; Riolo, P.; Ruschioni, S.; Isidoro, N.; Loreto, N.; et al. The bacterial biota of laboratory-reared edible mealworms (Tenebrio molitor, L.): From feed to frass. *Int. J Food Microbiol.* **2018**, *272*, 49–60. [CrossRef] [PubMed]
10. Nugaeva, N.; Gfeller, K.Y.; Backmann, N.; Lang, H.P.; Düggelin, M.; Hegner, M.; et al. Micromechanical cantilever array sensors for selective fungal immobilization and fast growth detection. *Biosens. Bioelectron.* **2006**, *21*, 849–856. [CrossRef]
11. Peterson, T.S.; Spitsbergen, J.M.; Feist, S.W.; Kent, M.L. Luna stain, an improved selective stain for detection of microsporidian spores in histologic sections. *Dis. Aquat. Org.* **2011**, *95*, 175–180. [CrossRef]
12. Morel, N.; Volland, H.; Dano, J.; Lamourette, P.; Sylvestre, P.; Mock, M.; Créminon, C. Fast and Sensitive Detection of Bacillus anthracis Spores by Immunoassay. *Appl. Environ. Microbiol.* **2012**, *78*, 6491–6498. [CrossRef]
13. Roy, M.; Jin, G.; Seo, D.; Nam, M.-H.; Seo, S. A simple and low-cost device performing blood cell counting based on lens-free shadow imaging technique. *Sens. Actuators B* **2014**, *201*, 321–328. [CrossRef]
14. Li, G.; Zhang, R.; Yang, N.; Yin, C.; Wei, M.; Zhang, Y.; Sun, J. An approach for cell viability online detection based on the characteristics of lensfree cell diffraction fingerprint. *Biosens. Bioelectron.* **2018**, *107*, 163–169. [CrossRef]
15. Roy, M.; Seo, D.; Oh, C.-H.; Nam, M.-H.; Kim, Y.J.; Seo, S. Low-cost telemedicine device performing cell and particle size measurement based on lens-free shadow imaging technology. *Biosens. Bioelectron.* **2015**, *67*, 715–723. [CrossRef]
16. Kwak, Y.H.; Lee, J.; Lee, J.; Kwak, S.H.; Oh, S.; Paek, S.-H.; Ha, U.-H.; Seo, S. A simple and low-cost biofilm quantification method using LED and CMOS image sensor. *J. Microbiol. Methods* **2014**, *107*, 150–156. [CrossRef]
17. Ishida, T.; Shimamoto, T.; Kaminaga, M.; Kuchimaru, T.; Kizaka-Kondoh, S.; Omata, T. Microfluidic High-Migratory Cell Collector Suppressing Artifacts Caused by Microstructures. *Micromachines* **2019**, *10*, 116. [CrossRef]
18. Yang, N.; Zhang, R.; Xu, P.; Xiang, Z. Optimised Microfluidic Device for Assessing A Rheumatoid Factor in Patients. *Electronics World* **2013**, *119*, 40–43.
19. Xu, J.; Zhang, J.; Wang, H.; Mi, J. Fine Particle Behavior in the Air Flow Past a Triangular Cylinder. *Aerosol Sci Tech.* **2013**, *47*, 875–884. [CrossRef]
20. Zhang, T.; Takahashi, H.; Hata, M.; Toriba, A.; Ikeda, T.; Otani, Y.; Furuuchi, M. Development of a Sharp-Cut Inertial Filter Combined with an Impactor. *Aerosol. Air Qual. Res.* **2017**, *17*, 644–652. [CrossRef]

21. Chung, A.J.; Gossett, D.R.; Carlo, D.D. Three Dimensional, Sheathless, and High-Throughput Microparticle Inertial Focusing Through Geometry-Induced Secondary Flows. *Small* **2013**, *9*, 685–690. [CrossRef]
22. Seo, J.; Lean, M.H.; Kole, A. Membrane-free microfiltration by asymmetric inertial migration. *Appl. Phys. Lett.* **2007**, *91*, 033901. [CrossRef]

 *micromachines*

Article

Detection of Cigarette Smoke Using a Surface-Acoustic-Wave Gas Sensor with Non-Polymer-Based Oxidized Hollow Mesoporous Carbon Nanospheres

Chi-Yung Cheng [1], Shih-Shien Huang [2], Chia-Min Yang [2], Kea-Tiong Tang [3] and Da-Jeng Yao [1,4,*]

[1]   Institute of NanoEngineering and MicroSystems, National Tsing Hua University, Hsinchu 30013, Taiwan; marybrian5498@gmail.com
[2]   Department of Chemistry, National Tsing Hua University, Hsinchu 30013, Taiwan; sean00132032@gmail.com (S.-S.H.); cmyang@mx.nthu.edu.tw (C.-M.Y.)
[3]   Department of Electrical Engineering, National Tsing Hua University, Hsinchu 30013, Taiwan; kttang@ee.nthu.edu.tw
[4]   Department of Power Mechanical Engineering, National Tsing Hua University, Hsinchu 30013, Taiwan
*   Correspondence: djyao@mx.nthu.edu.tw; Tel.: +886-3-5715131 (ext. 42850)

Received: 31 March 2019; Accepted: 22 April 2019; Published: 24 April 2019

Abstract: The objective of this research was to develop a surface-acoustic-wave (SAW) sensor of cigarette smoke to prevent tobacco hazards and to detect cigarette smoke in real time through the adsorption of an ambient tobacco marker. The SAW sensor was coated with oxidized hollow mesoporous carbon nanospheres (O-HMC) as a sensing material of a new type, which replaced a polymer. O-HMC were fabricated using nitric acid to form carboxyl groups on carbon frameworks. The modified conditions of O-HMC were analyzed with Scanning Electron Microscopy (SEM), Fourier transform infrared spectrometry (FTIR), and X-ray diffraction (XRD). The appropriately modified O-HMC are more sensitive than polyacrylic acid and hollow mesoporous carbon nanospheres (PAA-HMC), which is proven by normalization. This increases the sensitivity of a standard tobacco marker (3-ethenylpyridine, 3-EP) from 37.8 to 51.2 Hz/ppm and prevents the drawbacks of a polymer-based sensing material. On filtering particles above 1 μm and using tar to prevent tar adhesion, the SAW sensor detects cigarette smoke with sufficient sensitivity and satisfactory repeatability. Tests, showing satisfactory selectivity to the cigarette smoke marker (3-EP) with interfering gases CH_4, CO, and CO_2, show that CO and CO_2 have a negligible role during the detection of cigarette smoke.

Keywords: surface acoustic wave; second-hand smoke; 3-ethenylpyridine; oxidized hollow mesoporous carbon nanosphere

1. Introduction

Second-hand smoke (SHS), which is also called environmental tobacco smoke (ETS) or passive smoking, is a serious environmental pollution, composed of the main-stream smoke exhaled by active smokers and the side-stream smoke expelled from the lit tobacco product, such as a cigarette [1]. More than 4700 substances are now recognized as constituents of cigarette smoke [2]. More than 60 compounds are known to cause cancer in a human body [3], such as tar (including mutagenic and carcinogenic agents produced from incomplete combustion during smoking), tobacco-specific nitrosamines (TSNA), polonium-210, polycyclic aromatic hydrocarbons (PAH), etc. [4]. SHS is harmful for everybody, especially pregnant women and children. Statistical data reveal that people exposed to SHS have a greater risk of lung cancer (20–30%), oral cancer, asthma, chronic obstructive pulmonary disease (COPD), coronary heart disease (25%), and childhood illness [5,6]. A medical report states that, from 2010 to 2014 in the USA, smoking rates of adults fell from 43% (1965) to 18% today, but the number of deaths caused by smoking and exposure to SHS is still estimated to be about 480,000 per year [7].

In the complicated mixture of SHS, nicotine and 3-ethenylpyridine (3-vinylpyridine, 3-EP) are two important markers for sensing cigarette smoke because they are specific to tobacco and the vapor phase in SHS. 3-EP is a pyrolysis product of nicotine [8]. Even though the concentration of 3-EP in a smoke area is typically less than nicotine, the stable characteristics and smaller rate of surface absorption of 3-EP leads to it still being used as an SHS marker in various related research [9]. Most measurements of ambient nicotine are analyzed with a gas chromatograph - mass spectrometer (GC-MS) and typically have a prolonged sampling period of days because of the small effective rates of sampling [10,11]. At present, conductive polymer-based sensors have been designed to monitor a tobacco-specific vapor such as nicotine or 3-EP for real-time detection, but some drawbacks such as a lack of repeatability or a long response time must be improved [12,13].

Devices based on microelectromechanical systems (MEMS) have been widely used in many areas such as particle separation [14], chemical detection [15,16], biomedical detection [17], and microfluidics [18,19]. Surface-acoustic-wave (SAW) devices are easily manufactured with MEMS techniques and are used for gas sensing. The basic principle of a SAW sensor is that the coated interdigital transducer (IDT) stimulates a piezoelectric substrate with a stable electrical resonant frequency when a radio frequency (RF) signals in and converts it to propagate mechanical waves along the surface of the device, which is known as a surface acoustic wave. When the target is adsorbed by the sensing layer, the surface acoustic wave slows because of the mass loading. The resonant frequency decreases and the mass change would be proportional to the gas concentration observed as a frequency shift. Because the energy of SAW highly concentrates on the substrate surface [20], it is sensitive to any perturbation on the surface such as temperature, mass change, conductivity, and elasticity. These properties endow SAW devices with many advantages such as extremely high sensitivity, small size, a wireless sensing platform, and a low working temperature [15].

The sensitive coating of an SAW sensor is typically a polymer film with a suitable functional group, polarity, molecular geometry, and more. Some devices in the literature combine nanomaterials such as nanowire, carbon nanotube (CNT), graphene into the polymer for improved detection performance on increasing the sensing area, and leading to an increased mass loading [21–23]. However, a polymer-based sensing film has some drawbacks including lack of thermal stability and a mismatch between glass transition temperature and ambient temperature. Furthermore, some target vapor permeates into the polymer causing the polymer film to expand, called a swelling effect [24]. In this research, a surface-acoustic-wave (SAW) sensor is used to detect an SHS marker at a small concentration and cigarette smoke with sensing materials of two kinds – hollow mesoporous carbon nanospheres (HMC) combined with a polymer as a nanocomposite, and a modified nanomaterial without a polymer called oxidized hollow mesoporous carbon nanospheres (O-HMC). A sensitive and real-time cigarette sensor can help one avoid the SHS damage in public places such as a hospital and a school.

2. Materials and Methods

2.1. Preparation of a Surface-Acoustic-Wave (SAW) Sensor

The SAW chips with a delay line were fabricated with MEMS techniques on a 128° YX-lithium niobate (LiNbO3) piezoelectric substrate, which has a large electromechanical coupling coefficient (K2, 5.5%) and large velocity (3992 m/s) [25]. The delay line consisted of a pair of input and output interdigital transducers (IDT) with each having 50 pairs of fingers with gold (thickness 100 nm) and chromium (20 nm) deposited with an e-gun evaporator. The acoustic wavelength of the interdigital transducers (IDT) is designed to be about 34 μm. The operating frequency of the SAW device is about

114.2 MHz near 23 °C The insertion losses (IL) range from 6 to 9 dB, which is characterized with a network analyzer.

To complete the fabrication, the finished wafer was cut into chips. The dimensions of each device were 13.4 mm × 7.4 mm × 0.5 mm. To decrease the gas diffusion volume and to decrease the response time, a self-designed micro-chamber (800 µL) is mounted on a printed-circuit board (PCB). The gas is delivered with dispensing needles, which is shown in Figure 1a.

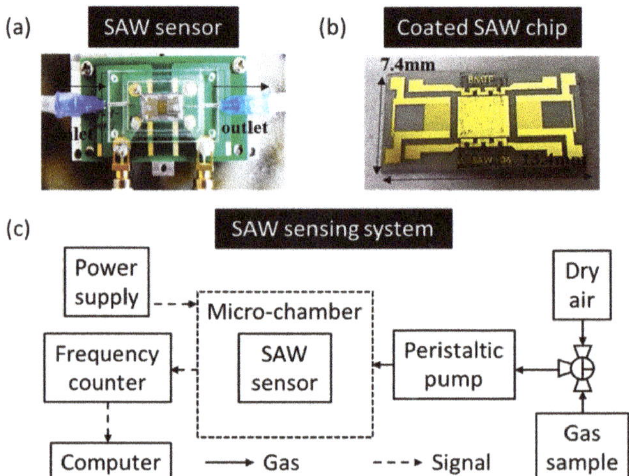

Figure 1. (**a**) Image of surface-acoustic-wave (SAW) sensor with a micro-chamber. (**b**) Spin-coated SAW chip with oxidized hollow mesoporous carbon nanospheres (O-HMC). (**c**) Schematic diagram of SAW sensing system.

2.2. Sensitive Materials

In this research, sensing materials of two types were chosen for the detection of cigarette smoke. One is a polymer-based and the other is only a nanostructure, without a polymer. For a polymer-based sensing film, poly-acrylic acid (PAA) is used for its universality. It has a glassy state near 23 °C because its glass transition temperature is about 106 °C, so it is hard to adsorb a target like a flexible polymer film [26]. To increase the reaction area and to decrease the response time, hollow mesoporous carbon nanospheres (HMC) are used to produce a nanocomposite sensing material (PAA-HMC). Each HMC has a specific surface area about 2350 m^2/g. The average size is in the range of 80 to 120 nm with a porosity of 4 nm.

To avoid the disadvantages of a polymer, a non-polymer-based sensing material, oxidized hollow mesoporous carbon nanospheres (O-HMC) is a suitable material for detection since it combines the properties of functional groups of a polymer (carboxyl group) and a large specific surface area of a nanomaterial. The fabrication of O-HMC is simple, modified by liquid-phase oxidation of HMC directly, and treated with nitric acid (2.5 M) at an appropriate temperature and for an appropriate period to introduce functional groups on the carbon framework [27]. These two sensing materials are solid after fabrication. In attempt to facilitate the coating, a sensing material (10 mg) is added to ethanol (1 mL) as a solvent. Before coating, the aggregation of this sensing material was prevented through ultra-sonication for 20 min, spin coating with great uniformity at 1500 rpm, and heating at 90 °C for 10 min to remove the solvent, which is shown in Figure 1b. The black dots on the delay line are the O-HMC.

2.3. Sensing System

The SAW sensing system is shown in Figure 1c. A standard gaseous SHS marker (3-EP) is prepared on dipping 3-ethenylpyridine liquid into a polytetrafluoroethene-based (PTFE) sampling bag to produce a saturated vapor. A three-way valve installed on the sampling bag can switch a sample or dry air from an air compressor (relative humidity about 20%). A sample of cigarette smoke is collected from a lit cigarette (Marlboro, tar 10 mg and nicotine 0.8 mg per cigarette) and placed on the bottom of a glass syringe. The cigarette smoke was drawn with a pump (Thomas, diaphragm pump 2002, 400 mL/min) into a sampling bag. Before detection of the cigarette smoke, the sampling bag was left at ambient temperature for 15 min to preclude any influence of temperature. For a dynamic detection, a gas sample was drawn into the micro-chamber with a peristaltic pump (flow rate of 20 mL/min). At the beginning of tests, dry air flowed into the chamber until the frequency response was stable. The gas sample channel was then switched for adsorption, and, eventually, the dry air valve was opened again to purge the sensing material and to complete one cycle. Since the temperature and humility effects by a piezoelectric subtract would be relatively large, the un-coated sensing chip would be used to eliminate the environmental interferences. All detection took an untreated HMC-coated SAW sensor as a reference for improved stability [15,16,28,29]. The frequency response was recorded in a computer with a general purpose interface bus (GPIB) card, which was connected to a frequency counter, and which monitored the frequency change between the sensing chip and the reference.

3. Results

3.1. Sensitive Analysis of a Material

The surface morphologies of HMC and O-HMC with varied chemically modified parameters were analyzed using a scanning electron microscope (SEM), as shown in Figure 2. The original HMC showed the size to range from 80 to 120 nm and its mesopores to have a diameter about 4 nm, which is shown in Figure 2a. After oxidation, some carbon species were cleaved from the carbon nanospheres, which made an incomplete spherical appearance shown in Figure 2b. When the modified conditions are too harsh, the carbon nanospheres are damaged with HNO_3, which leads to structural collapse to decrease the pore size. This might be explained by the repair of surface oxygen on the carbon walls [27]. The O-HMC seem to crosslink with each other, shown in Figure 2c, and show a serious aggregation problem that many O-HMC aggregate into clumps of diameter greater than 1 μm, which is shown in Figure 2d.

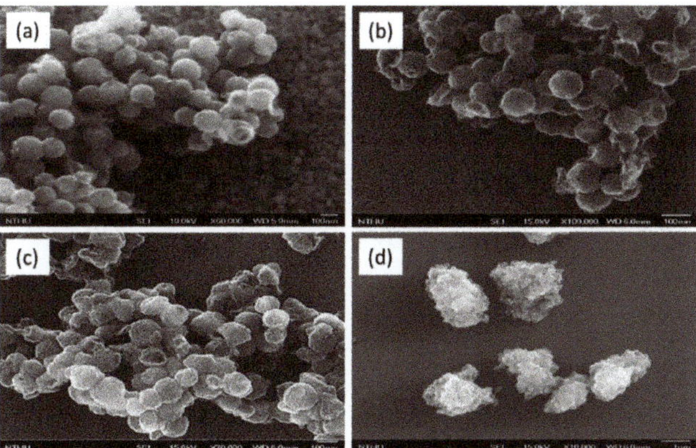

Figure 2. SEM images. (**a**) Hollow mesoporous carbon nanospheres (HMC). (**b**) O-HMC modified at 80 °C and 15 h. (**c**) O-HMC modified at 110 °C and 24 h. (**d**) O-HMC modified at 110 °C and 24 h on a small scale.

Considering the sensing application, the chemical structure of HMC and O-HMC was characterized with infrared spectra (FTIR) and scanned in a range from 500 to 4000 cm^{-1}, as shown in Figure 3. The HMC before and after oxidation seem to have a similar spectrum, but show a significant difference of the C=O vibrational line at 1730 cm^{-1}. A comparison with the spectrum before oxidation indicates that this signal denotes the presence of carbonyl or carboxyl groups. When HMC were treated with nitric acid, the stronger oxidation increased the intensity at 1730 cm^{-1}, which means that more carboxyl groups were modified.

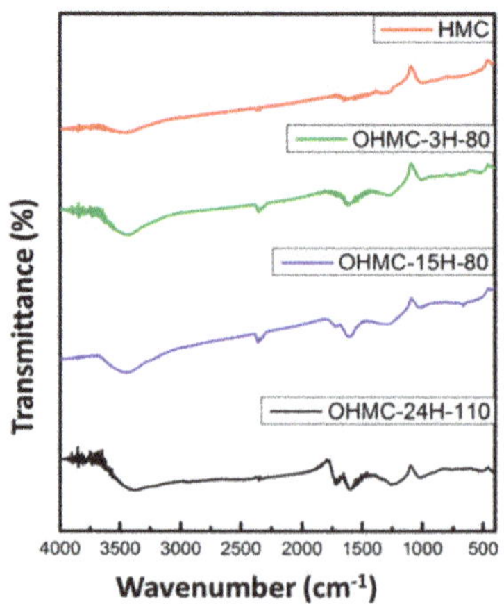

Figure 3. Infrared (IR) spectra of HMC and O-HMC with modified parameters.

The XRD pattern of the HMC shows three signals that are indexed as (110), (211), and (220) reflections of structure Ia3d, shown as Figure 4 [30]. A transformation of structure occurred after the removal of the silica template. The carbon framework became atomically disordered, as revealed by the (110) signal appearing. The (110) reflection is symmetrically forbidden for Ia3d [31]. Too much oxidation leads to structural changes, revealed by the weakened (110) and (211) XRD reflections by damaging the original structure or cleavage of the carbon. Based on the above analysis, to retain sufficient modification for great sensitivity and the complete nanostructures for a large surface area, O-HMC-80 °C-15 h was chosen as the appropriate sensing material for subsequent detection.

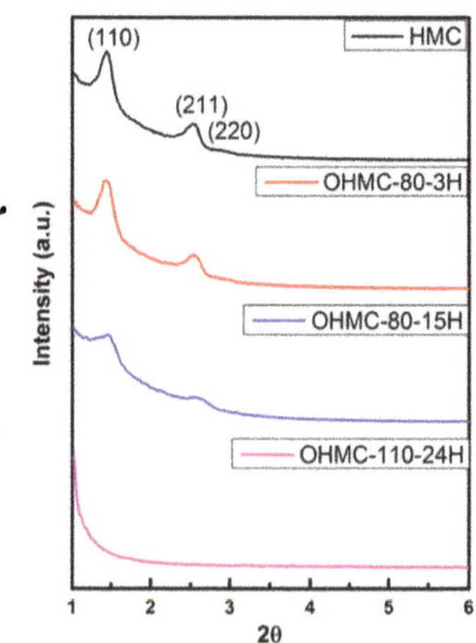

Figure 4. XRD of HMC and O-HMC with modified parameters.

3.2. Detection of the Standard Cigarette Marker

Figure 5 shows the frequency response of 3-EP detection (3 ppm) with sensing material of three types. It clearly shows that the sensing layer of O-HMC without a polymer is more sensitive than polymer-based PAA-HMC for detection of 3-EP, which is a greater frequency shift of 90 Hz. Taking 60 s for the adsorption, the desorption time is 90 s. Hydrogen bonding between the carboxylic acid and 3-EP become desorbed to the original frequency. Showing an effective reversible detection, this result proves that the modified O-HMC is a suitable choice to replace a polymer-based sensing material (PAA-HMC). In addition, the non-treated HMC shows slight adsorption, likely because some 3-EP is trapped in the porous nanostructure.

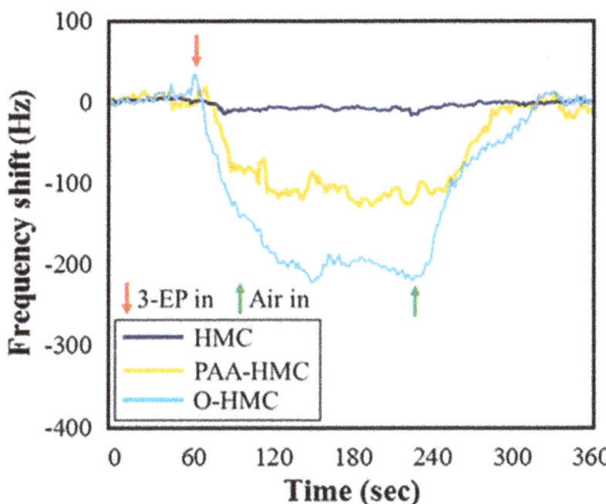

Figure 5. Frequency response of sensing materials for detection of 3-EP (3 ppm).

3.3. Normalization

The mass loading of the sensitive layer is an important factor for the sensor performance. In general, more sensing material coating can cause a greater frequency shift. However, too much sensing material results in small energy transmission. To compare the sensitivity of various sensing materials with no influence caused by the amount of coating, normalization is a necessary step. The frequency shift of a surface-acoustic-wave sensor is assumed to conform to Equation (1) for an acoustically thin and perfectly elastic thin film, k_1 and k_2 are piezoelectric material parameters, f_0 is the central frequency of the SAW device, h is the thickness of the sensing film, ρ is the density of the sensing film, m is the mass of adsorbed molecules, and A is the coated area [32]. Equation (2) is derived from Equation (1) because $m_{coating}$ and $\Delta f_{coating}$ are constant after coating. m_{gas} is linear with Δf_{gas}, which means that the frequency shift caused from an inconsistent amount of coating can be calibrated on dividing by $\Delta f_{coating}$. In this research, the normalized frequency shift is defined as a ratio of the frequency shift caused by detection and coating ($\Delta f_{gas}/\Delta f_{coating}$) [16].

$$\Delta f = (k1 + k2)f_0^2 h \rho = (k1 + k2)f_0^2 m/A \tag{1}$$

$$\Delta f_{gas} / \Delta f_{coating} = m_{gas}/m_{coating} \tag{2}$$

This frequency shift caused on coating is shown in Table 1. In the non-polymer-based sensing materials, it can observe that, when the chemical oxidation is stronger, the material surface is more hydrophilic because of more oxygen-containing functional groups introduced. The deposited amount after spin coating (1500 rpm, 30 s) remaining in the sensing area (gold) is greater, which results in a large frequency shift. The HMC has a hydrophobic property. Therefore, under the same conditions of spin, the coating remains less than O-HMC. The original frequency shift and normalized frequency shift with 3-EP at varied concentrations is shown in Figure 6. Both types of sensing materials present satisfactory linearity, and the sensitivity (defined as the slopes of the regression line, Δf_{gas}/sample concentration) of PAA-HMC and O-HMC are 37.8 and 51.1 Hz/ppm. The detection limit (LOD) of the sensor is less than 1 ppm. Furthermore, it is insufficiently accurate to analyze the sensitivity of sensing materials of the two types from the original frequency shift because the deposited O-HMC is much less than PAA-HMC. The frequency shifts of the coatings ($\Delta f_{coating}$) were 60 kHz for O-HMC and 120 kHz for PAA-HMC. After normalization, it shows a larger difference and the sensitivity can be more clearly compared for the two sensing materials. The slope of the regression line for O-HMC is about 2.7 times that for PAA-HMC.

Table 1. Surface-acoustic-wave (SAW) Resonant frequency before and after coating.

Sensing Materials	Before Coated (MHz)	After Coated (MHz)	Coated Frequency Shift (Hz)
Polymer based			
PAA-HMC	114.71	114.59	120,000
Non-polymer based			
O-HMC-110 °C-24 h	114.98	114.88	100,000
O-HMC-80 °C-15 h	114.12	114.07	50,000
O-HMC-80 °C-3 h	114.16	114.11	50,000
HMC	114.16	114.13	30,000

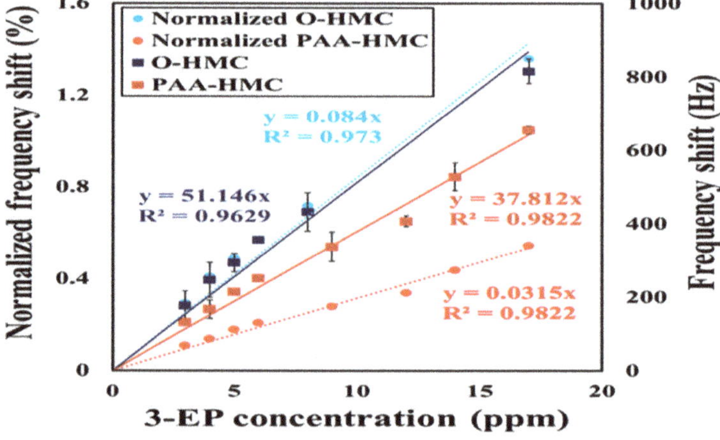

Figure 6. Frequency shift (deep blue and deep red) and normalized frequency shift (blue and red) of sensing materials O-HMC, polyacrylic acid, and hollow mesoporous carbon nanospheres (PAA-HMC).

3.4. Detection of Cigarette Smoke

Besides the detection of pure 3-ethenylpyridine, cigarette smoke is also detected with the same system and same sensing material. Unlike the pure compound of 3-EP, the cigarette smoke is a complicated mixture, including toxic volatile organic compounds (VOC) of many kinds such as benzene, toluene, formaldehyde, phenol, and sticky tar, which is produced by incomplete combustion. Figure 7a is the frequency response of cigarette smoke (burning 1 cigarette) detected with O-HMC, but the sample of cigarette smoke was collected without a filter. It shows that the frequency response continuously decreases when the cigarette smoke flows into the micro-chamber. The adsorption cannot achieve a dynamic balance and desorption cannot fully return to the original resonant frequency. This abnormal phenomenon of detection was likely due to the surface adhesion of tar. When the cigarette smoke sample was collected without a filter, sticky tar and particulate matter (PM) might cover the detection area, even sticking to the porous structure of the sensing material, which makes permanent mass loading and causes incomplete desorption.

Figure 7b is the frequency response for detecting cigarette smoke with O-HMC for five successive cycles. The sample of cigarette smoke (1 cigarette burning) is collected with a 1-μm filter. The SAW frequency decreased immediately when the cigarette smoke was introduced. Compared to the sample of cigarette smoke without a filter, the filtered cigarette smoke sample took about 4 min to complete the adsorption and 15 min to complete the desorption and to recover to the original frequency. The average frequency shift was about 4200 Hz. After filtering, it can also detect repeatedly like the detection of pure 3-ethenylpyridine, but unfiltered tar still slightly influences the sensor performance in repeatability.

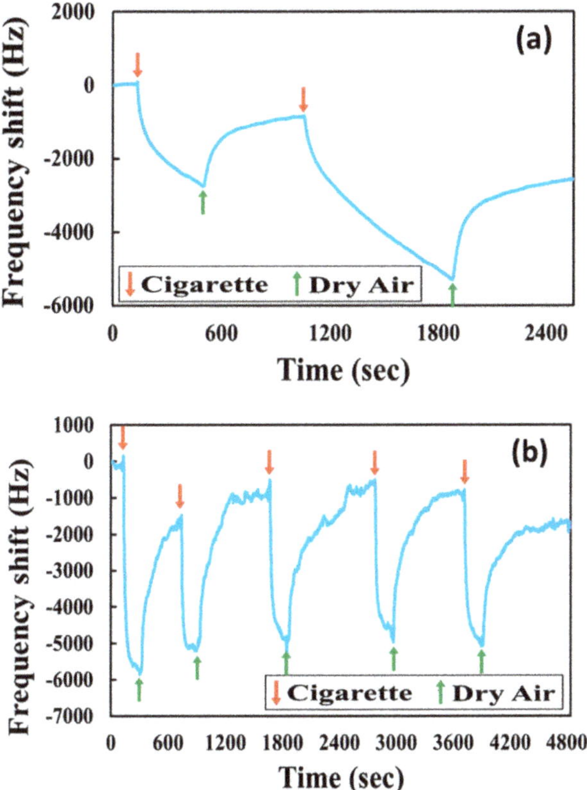

Figure 7. (**a**) Frequency response of the O-HMC-coated SAW sensor for detection of cigarette smoke without a filter. (**b**) Frequency response of the O-HMC-coated SAW sensor for detection of cigarette smoke with a 1-μm filter.

3.5. Selectivity of an Oxidized Hollow Mesoporous Carbon Nanospheres (O-HMC)-Coated SAW Sensor

When a cigarette burns, many chemical compounds are released into the air such as carbon oxide, carbon dioxide, methane, furans, PAH, nicotine, and 3-EP [4,13,33]. Because the properties of some compounds in cigarette smoke are similar to those of the cigarette marker and are also adsorbed on the sensing material, it is difficult to conclude that the total frequency shift in the detection of cigarette smoke is contributed from the cigarette marker. To determine the influence of non-cigarette- related substances, a selectivity test is necessary. Figure 8 shows a comparison of sensitivity of the SAW sensor coated with an O-HMC-sensitive material towards CO, CO_2, CH_4, and 3-EP (SHS marker), which indicates a high selectivity to the cigarette smoke marker because of the high affinity of 3-EP to

O-HMC. These results indicate that the frequency shift due to CO and CO_2 could be negligible during the detection of cigarette smoke, but CH_4 shows a larger adsorption to O-HMC as the micro-pores and meso-pores in the carbon of diameter less than 2 nm has an effective adsorption capacity for methane [34]. As the SAW sensor performance is also seriously affected by moisture, the frequency shift at varied relative humidity (RH) is also measured to calibrate the influence of water molecules. Overall, the selectivity of O-HMC toward cigarette smoke is adequate to apply the SAW sensor and detection under real conditions.

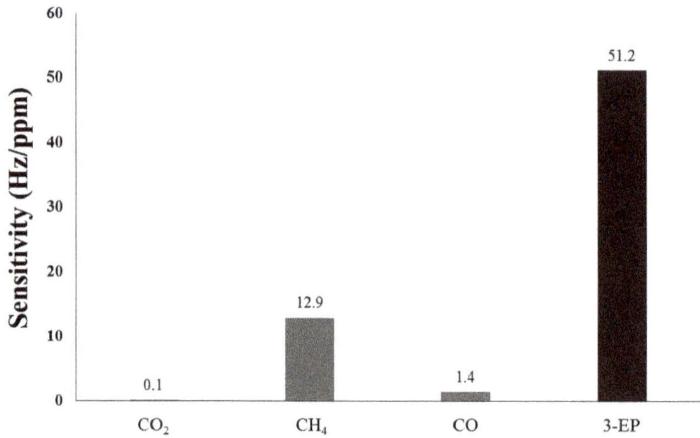

Figure 8. Selectivity of SAW sensor coated with O-HMC towards CO_2 (1.27%), CH_4 (150 ppm), CO (350 ppm), and 3-EP (standard cigarette marker, 17 ppm).

4. Conclusions

Statistical data indicate that persons exposed to second-hand smoke (SHS) have a greater risk of lung cancer and coronary heart disease. The surface-acoustic-wave (SAW) gas sensor coated with oxidized hollow mesoporous carbon nanospheres (O-HMC) detects the second-hand smoke marker (3-ethenylpyridine) and cigarette smoke. This non-polymer sensitive coating is made through treatment with nitric acid and has many carboxyl groups to bond with the tobacco marker. The O-HMC is more sensitive than PAA-HMC and can avoid the shortcomings of a polymer-based sensing material. It shows satisfactory selectivity to CO, CO_2, CH_4, and 3-EP. The large specific surface area caused by the porous structure and the self-designed micro-chamber lead to rapid detection at a small flow rate. Besides the detection of the pure compound, the SAW sensor detects cigarette smoke repeatedly with a filter of suitable size to remove excess tar and particles, which have great potential as a real-time smoke detector. This SAW sensor is an important distinction from passive sampling and time-consuming measurements of ambient nicotine. It allows for a demonstration of the concentration changes in the air cigarette marker from a specific smoking condition and provides the immediate protection from tobacco hazards.

Author Contributions: C.-M.Y., K.-T.T., and D.-J.Y. conceived and designed the experiments. C.-Y.C. and S.-S.H. performed the experiments. C.-Y.C., S.-S.H., and D.-J.Y. analyzed the data. C.-M.Y., K.-T.T., and D.-J.Y. prepared the manuscript.

Funding: This research was funded by Ministry of science and technology (MOST), Taiwan (grant number MOST 103-2220-E-007-023).

Acknowledgments: Thanks for Fan-Gang Tseng in National Tsing Hua University, provided equipment for fabrication of SAW sensing chips.

Conflicts of Interest: The authors declare no conflict of interest.

References

1. Eriksen, M.P.; LeMaistre, C.A.; Newell, G.R. Health hazards of passive smoking. *Ann. Rev. Public Health* **1988**, *9*, 47–70. [CrossRef] [PubMed]
2. Baker, R.R.; Pereira da Silva, J.R.; Smith, G. The effect of tobacco ingredients on smoke chemistry. Part I: Flavourings and additives. *Food Chem. Toxicol.* **2004**, *42*, 3–37. [CrossRef]
3. Baker, R.R.; Massey, E.D.; Smith, G. An overview of the effects of tobacco ingredients on smoke chemistry and toxicity. *Food Chem. Toxicol.* **2004**, *42*, S53–S83. [CrossRef]
4. Apelberg, B.J.; Hepp, L.M.; Avila-Tang, E.; Gundel, L.; Hammond, S.K.; Hovell, M.F.; Hyland, A.; Klepeis, N.E.; Madsen, C.C.; Navas-Acien, A.; et al. Environmental monitoring of secondhand smoke exposure. *Tob. Control* **2013**, *22*, 147–155. [CrossRef]
5. Hecht, S.S. Cigarette smoking: cancer risks, carcinogens, and mechanisms. *Langenbeck's Arch. Surg.* **2006**, *391*, 603–613. [CrossRef] [PubMed]
6. U.S. Department of Health and Human Services FDA Center for Drug Evaluation and Research; U.S. Department of Health and Human Services FDA Center for Biologics Evaluation and Research; U.S. Department of Health and Human Services FDA Center for Devices and Radiological Health. Guidance for industry: Patient-reported outcome measures: use in medical product development to support labeling claims: draft guidance. *Health Qual. Life Outcomes* **2006**, *4*, 79. [CrossRef]
7. U.S. Department of Health and Human Services. *The Health Consequences of Smoking—50 Years of Progress*; Report of the Surgeon General; U.S. Department of Health and Human Services: Rockville, MD, USA, 2014.
8. Vainiotalo, S.; Vaaranrinta, R.; Tornaeus, J.; Aremo, N.; Hase, T.; Peltonen, K. Passive monitoring method for 3-ethenylpyridine: a marker for environmental tobacco smoke. *Environ. Sci. Technol.* **2001**, *35*, 1818–1822. [CrossRef] [PubMed]
9. Koszowski, B.; Goniewicz, M.L.; Czogala, J.; Zymelka, A.; Sobczak, A. Simultaneous determination of nicotine and 3-vinylpyridine in single cigarette tobacco smoke and in indoor air using direct extraction to solid phase. *Int. J. Environ. Anal. Chem.* **2009**, *89*, 105–117. [CrossRef]
10. Hammond, S.K.; Leaderer, B.P. A diffusion monitor to measure exposure to passive smoking. *Environ. Sci. Technol.* **1987**, *21*, 494–497. [CrossRef] [PubMed]
11. Pandey, S.K.; Kim, K.-H. A review of environmental tobacco smoke and its determination in air. *TrAC Trends Anal. Chem.* **2010**, *29*, 804–819. [CrossRef]
12. Liu, Y.; Antwi-Boampong, S.; BelBruno, J.J.; Crane, M.A.; Tanski, S.E. Detection of secondhand cigarette smoke via nicotine using conductive polymer films. *Nicotine Tob. Res.* **2013**, *15*, 1511–1518. [CrossRef] [PubMed]
13. Chowdhury, D. Ni-coated polyaniline nanowire as chemical sensing material for cigarette smoke. *J. Phys. Chem. C* **2011**, *115*, 13554–13559. [CrossRef]
14. Chen, C.C.; Chen, Y.A.; Liu, Y.J.; Yao, D.J. A multilayer concentric filter device to diminish clogging for separation of particles and microalgae based on size. *Lab. Chip.* **2014**, *14*, 1459–1468. [CrossRef]
15. Hao, H.; Tang, K.T.; Ku, P.H.; Chao, J.S.; Li, C.H.; Yang, C.M.; Yao, D.J. Development of a portable electronic nose based on chemical surface acoustic wave array with multiplexed oscillator and readout electronics. *Sens. Actuators B Chem.* **2010**, *146*, 545–553. [CrossRef]
16. Ku, P.-H.; Hsiao, C.Y.; Chen, M.J.; Lin, T.H.; Li, Y.T.; Liu, S.C.; Tang, K.T.; Yao, D.J.; Yang, C.M. Polymer/ordered mesoporous carbon nanocomposite platelets as superior sensing materials for gas detection with surface acoustic wave devices. *Langmuir* **2012**, *28*, 11639–11645. [CrossRef] [PubMed]
17. Chen, C.H.; Lin, C.T.; Hsu, W.L.; Chang, Y.C.; Yeh, S.R.; Li, L.J.; Yao, D.J. A flexible hydrophilic-modified graphene microprobe for neural and cardiac recording. *Nanomedicine* **2013**, *9*, 600–604. [CrossRef] [PubMed]
18. Huang, H.-Y.; Huang, P.-W.; Yao, D.-J. Enhanced efficiency of sorting sperm motility utilizing a microfluidic chip. *Microsyst. Technol.* **2015**, *23*. [CrossRef]
19. Huang, H.Y.; Shen, H.H.; Tien, C.H.; Li, C.J.; Fan, S.K.; Liu, C.H.; Hsu, W.S.; Yao, D.J. Digital microfluidic dynamic culture of mammalian embryos on an electrowetting on dielectric (EWOD) chip. *PLoS ONE* **2015**, *10*, e0124196. [CrossRef] [PubMed]
20. Drafts, B. Acoustic wave technology sensors. *Microw. Theory Tech. IEEE Trans.* **2001**, *49*, 795–802. [CrossRef]

21. Sivaramakrishnan, S.; Rajamani, R.; Smith, C.S.; McGee, K.A.; Mann, K.R.; Yamashita, N. Carbon nanotube-coated surface acoustic wave sensor for carbon dioxide sensing. *Sens. Actuators B Chem.* **2008**, *132*, 296–304. [CrossRef]
22. Sayago, I.; Fernández, M.J.; Fontecha, J.L.; Horrillo, M.C.; Vera, C.; Obieta, I.; Bustero, I. New sensitive layers for surface acoustic wave gas sensors based on polymer and carbon nanotube composites. *Sens. Actuators B Chem.* **2012**, *175*, 67–72. [CrossRef]
23. Liu, S.; Sun, H.; Nagarajan, R.; Kumar, J.; Gu, Z.; Hwan Cho, J.; Kurup, P. Dynamic chemical vapor sensing with nanofibrous film based surface acoustic wave sensors. *Sens. Actuators A Phys.* **2011**, *167*, 8–13. [CrossRef]
24. Grate, J.W.; McGill, R.A. Dewetting effects on polymer-coated surface acoustic wave vapor sensors. *Anal. Chem.* **1995**, *67*, 4015–4019. [CrossRef]
25. Kao, K.; Cheng, C.; Chen, Y.; Lee, Y.-H. The characteristics of surface acoustic waves on AlN/LiNbO3 substrates. *Appl. Phys. A* **2003**, *76*, 1125–1127. [CrossRef]
26. Levit, N.; Pestov, D.; Tepper, G. High surface area polymer coatings for SAW-based chemical sensor applications. *Sens. Actuators B Chem.* **2002**, *82*, 241–249. [CrossRef]
27. Bazuła, P.A.; Lu, A.-H.; Nitz, J.-J.; Schüth, F. Surface and pore structure modification of ordered mesoporous carbons via a chemical oxidation approach. *Microporous Mesoporous Mater.* **2008**, *108*, 266–275. [CrossRef]
28. Cheng, C.-Y.; Huang, S.-S.; Yang, C.-M.; Tang, K.-T.; Yao, D.-J. Detection of third-hand smoke on clothing fibers with a surface acoustic wave gas sensor. *Biomicrofluidics* **2016**, *10*, 011907. [CrossRef]
29. Chiang, M.-C.; Hao, H.-C.; Yang, C.-M.; Yao, D.-J. Detection of hazardous vapors including mixtures in varied conditions using a surface-acoustic-wave device. *ECS J. Solid State Sci. Technol.* **2018**, *7*, Q3120–Q3125. [CrossRef]
30. Lai, N.; Chih, Y.L.; Pei, H.K.; Li, L.C.; Liao, K.W.; Lin, W.T.; Chia, M.Y. Hollow mesoporous Ia3d silica nanospheres with singleunit-cell-thick shell: Spontaneous formation and drug delivery application. *Nano Res.* **2014**, *7*, 1439–1448. [CrossRef]
31. Ryoo, R.; Joo, S.H.; Jun, S. Synthesis of highly ordered carbon molecular sieves via template-mediated structural transformation. *J. Phys. Chem. B* **1999**, *103*, 7743–7746. [CrossRef]
32. Wohltjen, H.; Dessy, R. Surface acoustic wave probe for chemical analysis. I. Introduction and instrument description. *Anal. Chem.* **1979**, *51*, 1458–1464. [CrossRef]
33. Tan, T.L.; Lebron, G.B. Determination of carbon dioxide, carbon monoxide, and methane concentrations in cigarette smoke by fourier transform infrared spectroscopy. *J. Chem. Educ.* **2012**, *89*, 383–386. [CrossRef]
34. Mosher, K.; He, J.; Liu, Y.; Rupp, E.; Wilcox, J. Molecular simulation of methane adsorption in micro- and mesoporous carbons with applications to coal and gas shale systems. *Int. J. Coal Geol.* **2013**, *109–110*, 36–44. [CrossRef]

Article

A Microfluidic Split-Flow Technology for Product Characterization in Continuous Low-Volume Nanoparticle Synthesis

Holger Bolze [1,2], **Peer Erfle** [2,3] **Juliane Riewe** [2,4], **Heike Bunjes** [2,4], **Andreas Dietzel** [2,3] **and Thomas P. Burg** [1,5,*]

1   Max Planck Institute for Biophysical Chemistry, Göttingen 37077, Germany; holger.bolze@mpibpc.mpg.de
2   Center of Pharmaceutical Engineering, Technische Universität Braunschweig,
    Braunschweig 38106, Germany; p.erfle@tu-bs.de (P.E.); j.riewe@tu-braunschweig.de (J.R.);
    heike.bunjes@tu-braunschweig.de (H.B.); a.dietzel@tu-braunschweig.de (A.D.)
3   Institute of Microtechnology, Technische Universität Braunschweig, Braunschweig 38124, Germany
4   Institut für Pharmazeutische Technologie, Technische Universität Braunschweig,
    Braunschweig 38106, Germany
5   Department of Electrical Engineering and Information Technology, Technische Universität Darmstadt,
    64283 Darmstadt, Germany
*   Correspondence: tburg@mpibpc.mpg.de

Received: 29 January 2019; Accepted: 5 March 2019; Published: 9 March 2019

Abstract: A key aspect of microfluidic processes is their ability to perform chemical reactions in small volumes under continuous flow. However, a continuous process requires stable reagent flow over a prolonged period. This can be challenging in microfluidic systems, as bubbles or particles easily block or alter the flow. Online analysis of the product stream can alleviate this problem by providing a feedback signal. When this signal exceeds a pre-defined range, the process can be re-adjusted or interrupted to prevent contamination. Here we demonstrate the feasibility of this concept by implementing a microfluidic detector downstream of a segmented-flow system for the synthesis of lipid nanoparticles. To match the flow rate through the detector to the measurement bandwidth independent of the synthesis requirements, a small stream is sidelined from the original product stream and routed through a measuring channel with 2 × 2 μm cross-section. The small size of the measuring channel prevents the entry of air plugs, which are inherent to our segmented flow synthesis device. Nanoparticles passing through the small channel were detected and characterized by quantitative fluorescence measurements. With this setup, we were able to count single nanoparticles. This way, we were able to detect changes in the particle synthesis affecting the size, concentration, or velocity of the particles in suspension. We envision that the flow-splitting scheme demonstrated here can be transferred to detection methods other than fluorescence for continuous monitoring and feedback control of microfluidic nanoparticle synthesis.

Keywords: lipid nanoparticles; online analysis; microfluidics; plug flow mixer; fluorescence; precipitation; single particle analysis; nanoparticle characterization

---

1. Introduction

Continuous processes provide numerous advantages over batch processes for the production of chemical and pharmaceutical products [1–3]. For example, continuous processes often result in significantly lower waste production [4] and better process control than batch processes [5]. Importantly, continuous processes are often amenable to a microfluidic implementation, enabling efficient production of small batches [2,3]. Several microfluidic systems for the continuous production of nanoparticles have been demonstrated in the past. Examples include systems for the synthesis of quantum dots [6,7], metal particles [8–10], metal oxide particles [11], drug nanoparticles [12], and polymer nanoparticles for medical application [13,14]. Liposomes [15] and lipid nanoparticles [16] have raised great interest as drug carriers. Lipid nanoparticles are of special interest for encapsulating poorly water-soluble active ingredients to facilitate their transport in the bloodstream and uptake into cells [17,18]. Since—especially at early stages of drug discovery and development—drugs are often produced only in small amounts, there is an interest in continuous microfluidic processes for synthesizing lipid nanoparticles continuously and efficiently at a small scale. By this motivation, the synthesis of lipid particles in microsystems has been demonstrated previously [19–24].

Process stability over several hours or days is an important challenge for the synthesis of uniformly sized nanoparticles in microfluidic systems. During the long time of production, the process can be altered by time-dependent changes, including mechanical wear of the pumping system, a change in the reactant composition, or the blockage of microfluidic channels by particles and gas bubbles. Such changes can compromise the quality or make the product unusable. Therefore, being able to monitor the process output is of great interest for designing reliable and robust microfluidic systems for the synthesis of nanoparticles. In contrast to an end-point analysis of small batches, online analysis allows the continuous measurement of important product characteristics and the ability to counteract any change. Adding such an analysis to a microfluidic system is tied to special requirements. First, a low volume without large cavities or dead volume is critical to avoid back mixing and to minimize the time delay between synthesis and characterization. Second, the flow rate through the detector needs to be matched to the detection bandwidth. This is important in order to ensure that particles pass at an adequate frequency while spending sufficient time in the detector to elicit a quantifiable response.

To analyze particles in a microfluidic setup, different kinds of detectors have been proposed [25,26]. The most commonly used techniques for analyzing nanoparticles are bulk methods, which analyze the collective effect induced by all particles in a reference volume. Examples of such techniques include dynamic light scattering [27,28], fluorescence spectroscopy [7,29], ultraviolet–visible (UV/Vis) spectroscopy [30], and X-ray absorption [31,32]. The drawback of bulk analysis techniques is a limited capability to detect samples with high polydispersity or a multimodal distribution. Another way to detect particles is by single particle analysis (SPA). Among the most well-known SPA-based systems are flow cytometers and cell sorters [33]. SPA detectors in microfluidic systems can use different detection principles like impedance detection [34–38], static light scattering [39–43], and fluorescence-based detection [43–48]. Often two or three of these detection principles are coupled [34,43,47,49,50]. Most microfluidic single-particle detectors described in the literature have been developed to detect and differentiate particles bigger than 1 μm. Smaller particles are significantly more difficult to detect, as the signal strength in all the above methods decreases nonlinearly with size [35,36,40–44,46,48,51,52].

Therefore, to achieve an adequate signal-to-noise ratio, it is important to choose the flow rate through the detector slow enough to be able to measure at a relatively low bandwidth. However, the direct sequential coupling of continuous-flow synthesis with a flow-through detector does not allow independent flow-rate adjustment. Therefore, while nanoparticles have been measured in microfluidic systems by different bulk sensors online [6,7,27,28,30–32], there are no studies, to our knowledge, combining a low sample volume microfluidic online SPA detector for sub-micron particles with a nanoparticle synthesis setup.

In this study, we combine a microfluidic lipid nanoparticle precipitation setup with a fluorescence detection system able to detect the nanoparticles in a small stream. In our system, lipid nanoparticles are synthesized by solvent-antisolvent-precipitation, in which an organic phase containing the lipid and a fluorescent dye is mixed with an excess of water [53]. The organic solvent, which is miscible with water, diffuses into the much bigger water volume. In the mixture, the character of the water dominates so that the lipids precipitate, encapsulating the dye. Since a higher oversaturation leads to a reduced particle size, rapid mixing to homogeneity is important to avoid concentration gradients that would result in a broad size distribution. To achieve this, we employed a plug-flow mixer first described by Erfle et al. [24], in which the fluids that are to be mixed are combined in plugs divided by gas. As these plugs propagate through the system, rapid mixing takes place due to recirculation of the fluids within the plugs.

The detector uses a micron-sized channel instead of a focusing sheath stream to ensure the measurement of single particles in a defined detection spot [48,51,54]. The whole particle suspension is flown through a microfluidic channel. A much smaller measurement channel branches off this channel. The small profile of the measurement channel (2 μm width and height) confines the volume of liquid and ensures that a great majority of the detection events is caused by isolated single particles. Detection itself is accomplished by focusing a laser on a part of the small channel to excite the fluorescent dye Nile Red, which is added to the lipid nanoparticles during precipitation. A characteristic phenomenon of Nile Red is that its fluorescence is heavily quenched in water so that the accumulated lipophilic dye molecules in the lipid particles are much brighter than the ones remaining in water [55]. By the combination of the quenching and the lipophilic accumulation in the lipid particles, the particles generate a fluorescent intensity much stronger than the surrounding liquid. The fluorescence emission is collected and analyzed. Here each passing particle generates a peak, the height of which corresponds to the number of fluorophores in the particle. The length of the peaks corresponds to the residence time in the detection spot. Assuming homogenous distribution of fluorophores in the lipid volume, the peak height can be used as a measure for particle size, while the residence time allows the measurement of the velocity in the channel. Based on the velocity, geometrical parameters and the number of particles per minute, the concentration of particles can be measured.

2. Materials and Methods

2.1. Mixing Chip

The mixing system included a micromixer which operated according to the principle of segmented gas–liquid flow [24]. Borosilicate glass (Schott BOROFLOAT® 33, Schott AG, Mainz, Germany) was chosen as the material because it is mechanically and chemically stable, biocompatible, and transparent for optical observations [12,56,57]. The fabrication of the microsystem and the structuring of the glass was performed by femtosecond laser ablation (microSTRUCT c; 3D Micromac AG, Chemnitz, Germany). Details about the manufacturing process can be found in Erfle et al. [24]. The micromixer has a symmetrical design in which the solutions are divided after entry and reunited at the flow-focusing point (see Appendix A). By this arrangement, the organic phase was focused in the middle of the channel, and the contact to the wall was limited to prevent fouling. The streams were generated by two syringe pumps (neMESYS 290N, Cetoni GmbH, Korbussen, Germany). Due to the small dimensions, mixing by convection was limited. To increase the mixing efficiency, nitrogen was injected at the flow-focusing point where the streams merge, achieving immediate segmentation of the continuous flow into liquid plugs. In these liquid plugs, recirculation was induced, which increases the mixing rate. In the mixture, the oversaturation of solvent led to the precipitation of particles, so a higher oversaturation increased the number of small particles. Thus, better mixing quality led to a smaller and more homogeneous size of the precipitated particles. The gas stream was regulated by a pressure regulator (OB1 MK3, ELVESYS, Paris, France), which allowed to control the plug length and thereby influence the particle size. A camera was used to measure the plug size and allowed adjustment of the particle/plug size.

The mixing chip was linked to the detector by capillary tubing made of polyether ether ketone (PEEK) with an inner diameter of 0.5 mm and an outer diameter of 0.821 mm with a length between 5 and 10 cm (Figure 1). The major part of the product stream was routed to a pressurized vial without passing the detector. A fraction of the product stream was rerouted by a Y-connector to the detector. The vial was pressurized up to 0.8 bar to drive the liquid through the detector.

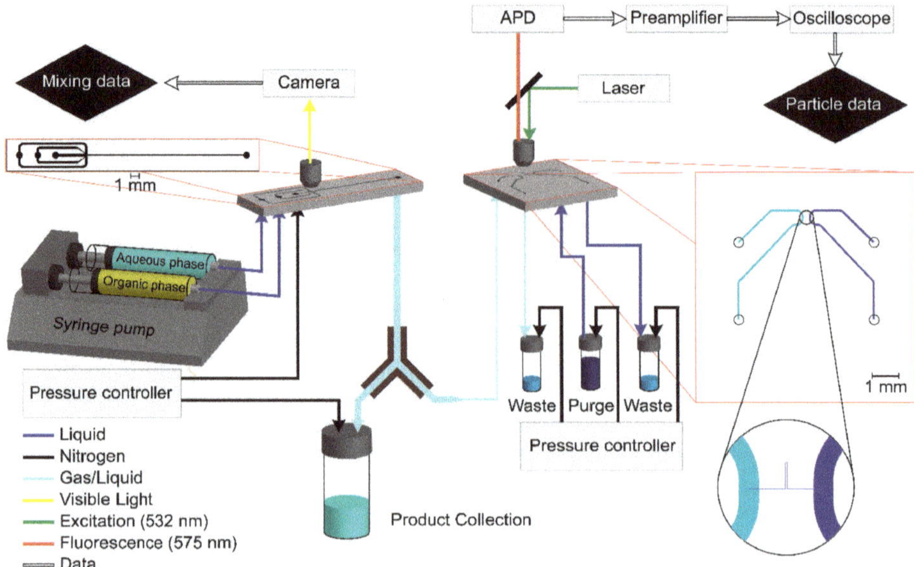

Figure 1. Flowchart and structure of the online analysis setup.

2.2. Detection Setup

The particle detection chip consisted of a structured silicon element closed by Borofloat glass. The chips were fabricated using lithography and standard silicon processing. The small measurement channel was first etched into silicon anisotropically by deep reactive ion etching (DRIE) on an STS Multiplex system (Surface Technology Systems, Newport, UK) using a Bosch process with alternating cycles of C4F8 for sidewall passivation and SF6 for etching. Large supply channels were then patterned by lithography and etched to a depth of 30 μm using a Bosch process. After this step, the channels were sealed by anodic bonding to an unstructured Borofloat wafer, and access holes were etched through the silicon from the backside using DRIE. The channel structure contained two 75 μm wide supply channels, which were connected by a measurement channel only 2 μm deep and wide. These dimensions were the design values. An example of a real device cross section is shown in the appendix.

The chip was connected to capillary tubing by a special holder allowing a microscope objective to operate at a distance less than 2 mm from the glass surface and, at the same time, connect the capillaries tightly to the backside of the chip. The buffer supply channel was filled with Hellmanex solution to purge stuck particles from the chip every four minutes. The outlet of the product supply channel was connected to a pressure-controlled vial. Additional pressure-controlled vials were connected to the inlet and outlet of the buffer supply channel. These pressure-controlled vials provided a tunable and stable flow, driving the particle suspension through the measurement channel or fresh buffer solution through the purge channel.

The analyzed channel area was illuminated by a 532 nm laser (CPS 532, Thorlabs, Newton, NJ, USA), which was focused on the chip by the microscope objective (LD-Plan NEOFLUAR 40×/0.6 Corr Ph2, Carl Zeiss AG, Oberkochen, Germany). The same objective collected the emitted light and projected it on a detector after passing a dichroic mirror (MD 568, Thorlabs, Newton, NJ, USA) and an emission filter (FELH0550, Thorlabs, Newton, NJ, USA). The light was analyzed by an avalanche photodiode (APD 110A, Thorlabs, Newton, NJ, USA). The signal of the photodiode was amplified 200 times and low-pass filtered at 10 kHz by a preamplifier (SR 560, Stanford Research Systems, Sunnyvale, CA, USA) before it was analyzed by an oscilloscope (PicoScope 4262, Cambridgeshire, UK). The oscilloscope measured the signal every 10 μs in a range of 5 V (DC).

For experiments using reference particles to characterize the whole system, the injected organic phase was replaced by an aqueous solution of reference particles, while all other solutions, pressures, and apparatus stayed the same to allow the characterization of parameters without changing concentration.

2.3. Chemicals and Solutions

The detector was calibrated and characterized using an aqueous solution containing fluorescent beads with a nominal diameter of 200 nm (F-8809, Thermo Fisher Scientific, Waltham, MA, USA) at a concentration of $4*10^6$ particles per milliliter. In addition, the solution contained Rhodamine to detect the flow direction even in case of clogging (10 µg/mL; Quality for Fluorescence, Merck, Darmstadt, Germany), sodium chloride (5.48 mg/mL, >99.8%, Carl Roth, Karlsruhe, Germany), sodium azide (20 µg/mL, >99%, Carl Roth, Karlsruhe, Germany), and sodium dodecyl sulfate (0.1 mg/mL, >99%, Merck, Darmstadt, Germany).

For the precipitation experiments, castor oil (5 mg/mL, Henry Lamotte Oils, Bremen, Germany), polysorbate 80 (2.5 mg/mL, BioXtra, Merck, Darmstadt, Germany), and Nile Red (8 µg/mL, technical grade, Merck, Darmstadt, Germany) were dissolved in ethanol and filtered through a 200 nm syringe filter (Polypropylene, VWR International, Radnor, PA, USA). The aqueous solution used consisted of deionized water filtered through a 200 nm syringe filter.

The system was purged with a 2% solution of Hellmanex (Hellmanex III, Helma Analytics, Müllheim, Germany) in deionized water, which was filtered through a 200 nm syringe filter (Polypropylene, VWR International, Radnor, PA, USA).

The gas bubble generation and pressurizing of vials was done with nitrogen.

2.4. Nanoparticle Characterization and Data Processing

To get additional information about the experiments, the produced particle suspensions were collected and analyzed by a NanoSight NS300 (Malvern Instruments, Malvern, UK) to measure the particle size distribution and concentration. For this measurement, the sample was diluted one thousand times, kept at 25 °C, and illuminated by a laser with a wavelength of 405 nm. The movement of particles was tracked for one minute with a rate of 25 frames per second. Assuming that the viscosity of the liquid corresponds approximately to that of water, the sample was analyzed with a detection limit of 20 arbitrary units. The detected particles were counted to measure the concentration and tracked to measure the Brownian motion, allowing the determination of the size of a single particle.

The data collected by the online detector were analyzed using a Matlab script, which set the baseline to zero by subtracting a low pass filtered version of the signal from itself. The new data track contained the peaks, which exhibited a plateau consisting of single elevated points. Our peak-finding algorithm marked only the highest value of each peak. After identifying the peaks, the script measured the number of peaks per second, the height of the peaks above the baseline, and the width of the peaks at the half height. A fixed value was added to the arbitrary units to compensate an offset induced by the baseline subtraction.

2.5. Experiments

Five different nitrogen plug sizes were used in the segmented-flow system to precipitate nanoparticles with different size distributions and concentrations. The liquids were mixed using controlled flow rates of 140 µL/min in the aqueous channel and 60 µL/min in the organic channel. The nitrogen pressure was adjusted to get the aimed plug size. From previous experiments, we knew that the mean particle size generated in plugs of the selected lengths was expected to vary between 95 nm and 170 nm in diameter [24].

The detector was started 5 min after a stable stream was established to allow time for the suspension to reach the detector. The detector measured the particles at 4 min intervals to allow purging of the measurement channel. After three cycles, the measurement was finished and the system was purged to continue with a different plug size.

3. Results and Discussion

3.1. Particle Concentration

First, we investigated the ability to measure the concentration of precipitated nanoparticles continuously using our coupled plug-flow mixer and fluidic bypass detector. The data (Figure 2a) show that by increasing the pressure drop in the synthesis, the concentration of particles in the product stream dropped. Importantly, this drop was consistently measured by our online detection method and off-line by batch measurements with the NanoSight instrument. For the NanoSight measurements, the sample flowing through the bypass was collected, and the concentration of nanoparticles was measured by counting the particles in the volume defined by the objective used in the NanoSight. The concentration of particles measured in the collected product showed the same dependence on the pressure drop as the microfluidic online detection. To find out whether the change in the number of detected particles was related to the change in precipitation or to varied flow velocity, the system was tested with reference particles under the same conditions. In comparison to the precipitated lipid nanoparticles, the reference particles showed no significant change in the number of detected particles with varied flow velocity. The change in particle number was therefore related to the synthesis in the chip and did not depend on changes in flow rates in the setup, which allowed us to correlate the microfluidic measured parameter with the concentration and thus calibrate these parameters for further processes (Figure 2b).

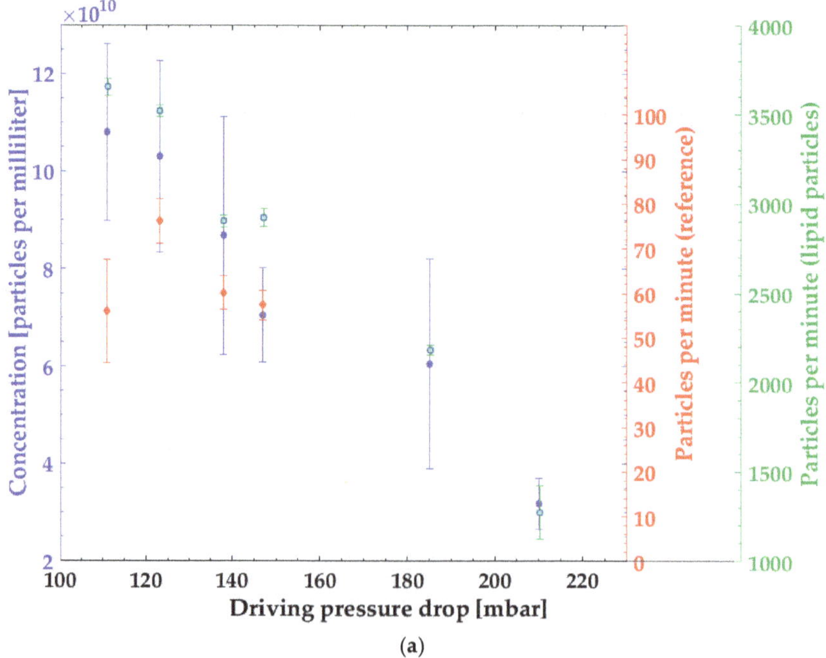

(a)

Figure 2. *Cont.*

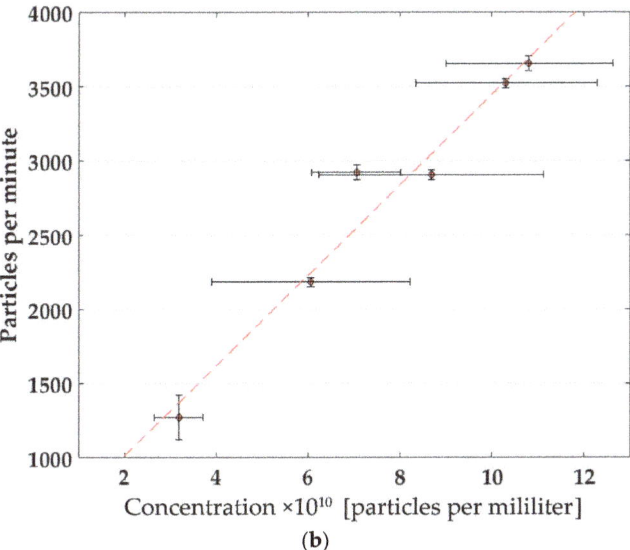

(b)

Figure 2. (**a**) Particles per minute measured with the microfluidic detector (red and green) and concentration of lipid nanoparticles in the collected product stream measured by the NanoSight system (blue). Error bars represent the standard errors of the respective measurement techniques. (**b**) Correlation of the microfluidic measured particles per minute and concentration.

3.2. Single-Particle Analysis by Fluorescence

Given that the concentration of Nile Red in the lipid nanoparticles is uniform, the maximum peak intensities that are detected in our setup are proportional to the volume of the particles. Histograms of the maximum peak intensities measured in experiments with different mixing conditions are shown in Figure 3a. Different mixing conditions were set up by varying the length of the gas plugs between the liquid segments in the mixer.

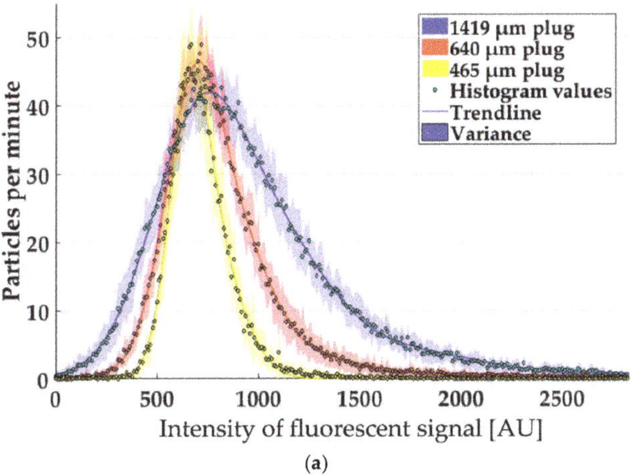

(a)

Figure 3. Cont.

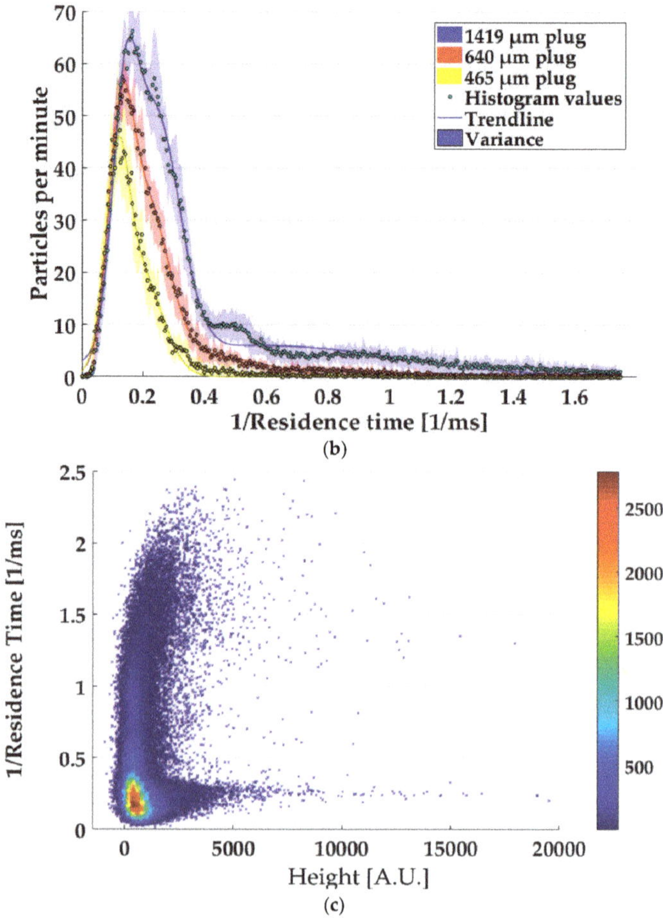

Figure 3. (**a**) Histograms of the fluorescence intensity of different experiments. (**b**) Histograms of the reciprocal residence time of different experiments. (**c**) Heatmap for all particles of a 1419 μm plug experiment to detect correlations between fluorescence intensity and residence time (colored areas in (**a**) and (**b**) represent the variance of three to four repeated experiments under the same conditions).

The histograms of the maximum peak intensities reveal a distribution which can be approximated by a superposition of Gaussian curves. The reduction in plug size leads to a reduction of the mean particle size up to 29%. This reduction is not due to a shift of the maxima in the histograms, but rather due to a narrowing of the distribution, which is more pronounced for particles much bigger than the mean. This change indicates that a bigger plug size leads to an increased polydispersity of the particle suspension.

The histograms of the reciprocal residence time show a very prominent peak in all experiments (Figure 3b). The height of the peaks varies between the experiments. These differences result from the different concentration of particles in each experiment as described before. A change of the mean value between the experiments is detectable, but is not due to a shift of the main peak. Instead, the histograms reveal additional peaks resulting from particles with lower residence times. These peaks are more prominent at larger plug sizes (and a lower pressure drop in the measurement channel) and lead to an increase of the mean value. Since the value represents the reciprocal residence time, this means that some particles were much faster, passing the measurement channel at conditions with a lower pressure drop.

The combination of fluorescence peak amplitude and width reveals very interesting details about the particle distribution, as shown in the scatter plot in Figure 4c. While most of the particles follow a narrow size distribution, some particles were scattered over a wider range of sizes and velocities. The interesting point is that the majority of the outliers seemed only to differ in one value from the majority of the points (Figure 3c). This independence of the outliers supports the hypothesis that the phenomena are not connected with each other and even very big particles can move freely in the channel. At a qualitative level, the scatter plot did not reveal any strong indication that outliers in size systematically deviate from the residence time distribution of the majority of the particles. This would be commensurate with the hypothesis that their velocity distribution follows the flow velocity field within the narrow channel. However, a more detailed quantitative analysis would be needed in the future to thoroughly elucidate potential systematic connections between residence time and particle size. Besides the predictable effects of steric hindrance within the small channel, connections between size and residence time distribution could also be suspected for reasons of stiction and other size/morphology-related effects.

To investigate whether the fluorescence intensity histograms measured with the microfluidic device were consistent with nanoparticle tracking analysis (NTA), the sample was collected at the outlet and analyzed with the NanoSight instrument. In this way, it was possible to measure the same sample in the microfluidic detector and by NTA and compare the resulting histograms. In this work, the focus was on measuring relative changes in a given distribution rather than absolute particle sizes. Therefore, the calibration was done by adjusting the arbitrary units on the *x*-axis of the histogram of particle diameters (measured by NTA) and the histogram of the third root of fluorescence intensities (measured by the microfluidic detector). The scale factor for the abscissa was chosen to obtain an overlap at the halved maxima of the two histograms (Figure 4a). The alignment was then transferred on two other measurements (Figure 4b,c) showing that a small shift of the peak maximum between the samples was traceable with both measurement techniques. This made it possible to trace a change in particle size in fluorescence and to establish a connection between particle size and fluorescence intensity. Note that although we consider this calibration sufficient for simple process monitoring, it would be interesting to pursue the option of an absolute independent calibration in future work. This could be accomplished by identifying suitable reference particles whose photophysical characteristics match those of the Nile Red-dyed lipid nanoparticles.

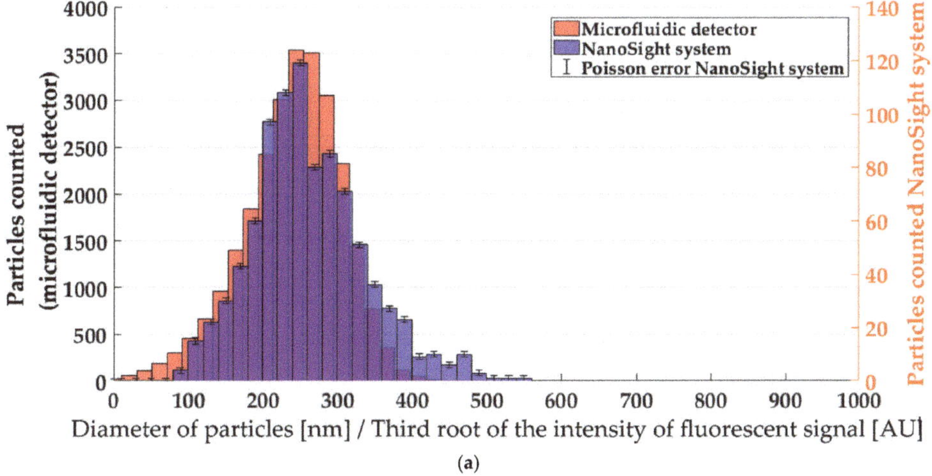

(**a**)

Figure 4. *Cont.*

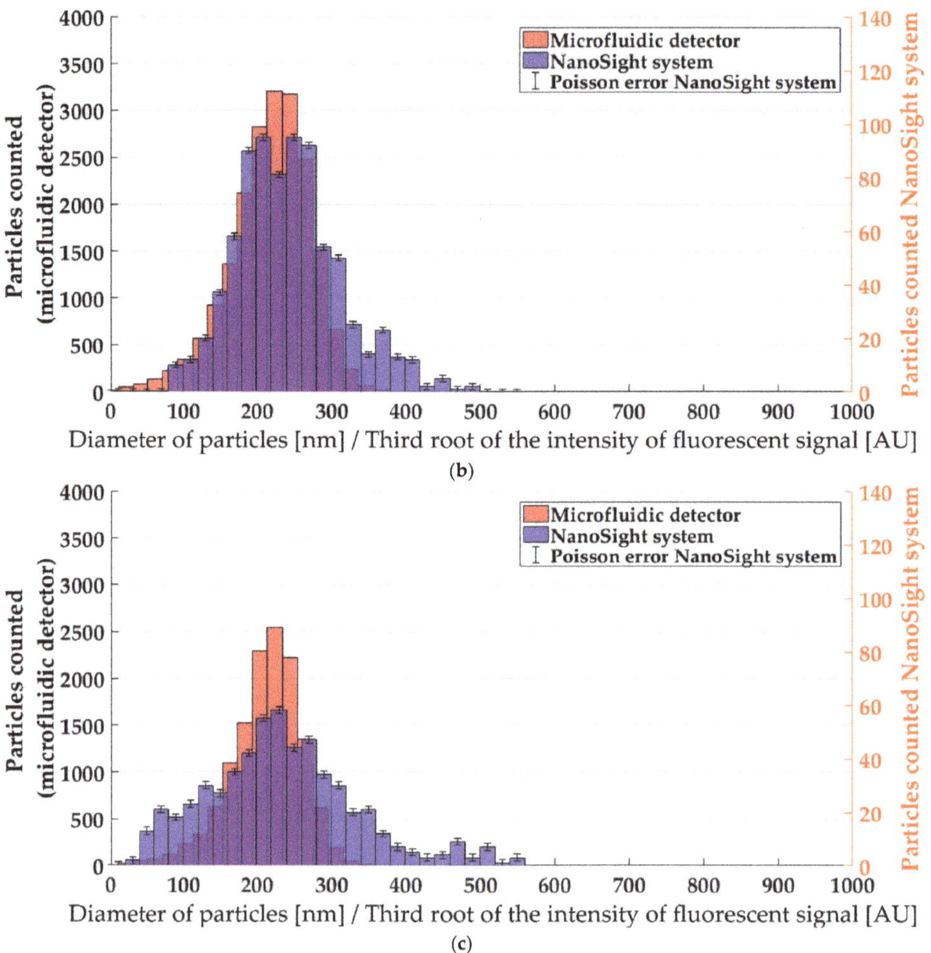

Figure 4. Comparison of qualitative histograms of the third root of fluorescence intensity and particle diameter at different precipitation conditions. The Poisson errors for the microfluidic detector were below a reasonable graphic expression in this context. (**a**) 812 µm plug size. (**b**) 529 µm plug size. (**c**) 386 µm plug size.

While the peak maximum and the reduction of particle concentration was traceable with both techniques, the polydispersity measured by the NanoSight system was increased in the second and third example (Figure 4b,c). A possible explanation for this is the much smaller number of particles measured with the NanoSight, combined with the possibility of systematic changes in the sample upon collection and transfer to the instrument. Further studies would be needed, however, to investigate this difference in more detail.

3.3. Continuous Monitoring to Detect Alterations in Nanoparticle Synthesis

The previous experiments demonstrated the capacity of the setup to detect fluorescent nanoparticles and recognize changes in the process of particle generation, but only under stable process conditions. As a next step, we studied the ability to detect transient situations when the process is destabilized at examples recorded during prolonged experiments previously described.

We monitored the signal from the fluorescence detector continuously and analyzed the three measured parameters (single-particle peak amplitude, peak frequency, and residence time) in 15 s intervals. As seen in Figure 5, different types of perturbations could be recorded. The analysis of the three parameters that were recorded allowed a detailed analysis of the transient phenomena. In one experiment, a slow drift in the particle-per-minute count and in the residence time per particle could be detected, as seen in the comparison between Figure 5a,b. This perturbation means that, despite constant synthesis condition, the throughput and velocity in the detector were reduced, which could be the result of fouling. Similarly, a much faster and unstable change in the residence time and the number of particles was detected (Figure 5d), which could be interpreted as clogging of the measurement channel. In both incidents recorded before, the intensity of the particles was not affected by the perturbations, indicating that the precipitation was unaffected. In contrast, in a third detected incident, the precipitation was disrupted, showing changes in all three parameters (Figure 5c). Since the concentration and velocity rose suddenly, a subtle change in flow rate and/or pressure within the fluidic system is likely the cause. A sudden pressure increase can lead to precipitation of larger particles, as seen in the intensity measurement. Using this method, we were able to detect these changes and record the time at which they occurred.

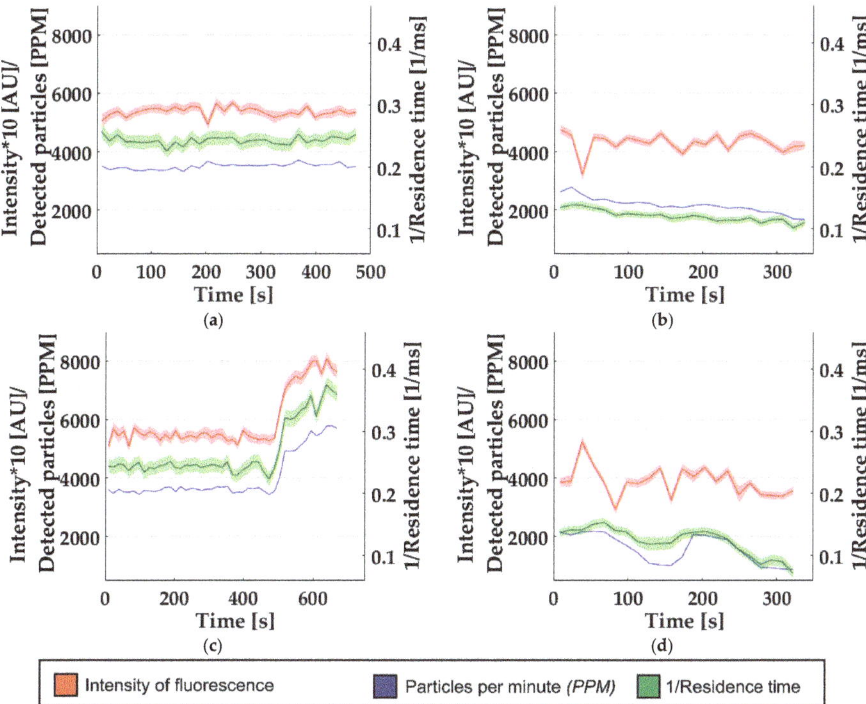

Figure 5. Detected perturbations of the online synthesis. Colored areas represent the standard error of the average value for 15 s measurement frames. (**a**) No incidents. (**b**) Drifting change. (**c**) Incident in synthesis. (**d**) Clogging.

In the future, this information could be used either as a feedback signal to stabilize the synthesis conditions or to suspend the process in time to prevent contamination of the collected product. In such applications, the response time of the detector is an important figure. Our two-chip synthesis/detector system provided interesting insights into realistically attainable response times with similar designs. Here the response time was limited mainly by two factors. First, there was a delay of 90 s in our system until the liquid reaches the detector. Second, at least two data points needed to be collected to identify a step change in the measured parameters. In our example, this added up to a minimum response time of 120 s until a change in synthesis conditions was visible. There is much room for improvement, however, for example by placing the synthesis and detection systems in closer proximity or by increasing the fluidic throughput. Our experience suggests that future systems could realistically be optimized to response times shorter than ten seconds.

4. Conclusions

By combining a microfluidic single particle analysis with a particle synthesis chip, we were able to realize an online analysis of nanoparticles synthesized in a microfluidic system. A unique characteristic of our detector is that only a small fraction of the generated particles is withdrawn from the product stream and guided through a detection channel, which has a volume of only 2 pL. Nanoparticles passing through the small channel one-by-one were detected by fluorescence. Transient changes in the particle synthesis were detected in different parameters and could be assigned to a specific time of the process. The single particle signals allow a very detailed analysis based on the histograms and the discussion of outliers.

By detecting the differences between different process conditions, we validated that the stability in size and concentration of the particles—in this process—can be detected, and that incidents happening during the process can be monitored.

To reach a feedback-controlled process, the analysis has to be automated in a single program to allow the fast detection of changes and to counteract those changes by adjusting control parameters of the synthesis (e.g., the driving pressures). Another valuable addition would be a robotic sample collection device that is reached by the stream later than the detector and so could respond to incidents by rerouting the tainted product and protect the already produced particle solution. Since a fast response of such a system is crucial, detection and synthesis should be combined in a single chip to lower the reaction time.

The comparison with particle size measurements, as shown in this work, allow a validation that the changes in intensity are related to a change in particle size. However, for a validated calibration, samples containing the same particles in a narrow size range would have to be measured with the microfluidic detector and an external calibration to correlate intensity to particle size.

In the future, the fundamental limits (e.g., minimum concentration, smallest possible response time) of this approach should also be further studied to allow a better understanding of the applicability of the detector. In addition, by adding static light scattering detection to the setup, the need for a fluorescent dye in the lipid nanoparticles could be eliminated to increase the number of applications for the detector. To increase the long-term stability of the detection, future setups could replace the small measurement channel with a wider channel that incorporates a flow focusing technique. This would increase the resistance of the method against clogging.

In summary, we developed a microfluidic device able to continuously monitor and detect changes in microfluidic nanoparticle synthesis processes, consuming only nanoliter samples.

Author Contributions: All authors contributed to designing the experiments. H.B. (Holger Bolze) and P.E. carried out the experiments. H.B. (Holger Bolze) and T.P.B. analyzed the data. The particle precipitation protocols were adapted from work of J.R. The precipitation chip was designed, built, and validated by P.E. The original draft was written by H.B. (Holger Bolze) and all authors contributed to the final manuscript.

Funding: Holger Bolze, Peer Erfle and Juliane Riewe acknowledge support by scholarships of the 'μ-Props—Processing of Poorly Soluble Drugs at Small Scale' program provided by the Niedersächsisches Ministerium für Wissenschaft und Kultur (MWK) of the federal state of Lower Saxony (Germany) and Boehringer Ingelheim, respectively. Heike Bunjes acknowledges support from the MWK of Lower Saxony (Germany) within the SMART BIOTECS alliance.

Acknowledgments: Holger Bolze thanks the Max Planck Institute for Biophysical Chemistry for hosting him during this project.

Conflicts of Interest: The authors declare no conflict of interest.

Appendix A

Mixing Chip

The mixing chip is built by combining two glass slides, which are both structured to a depth of 100 µm to achieve a round or oval profile [24]. The fluid channels branch in two symmetrical channels until all streams are combined in a crossing section (Figure A1). To ensure the purity of the liquids, the inlet channels contained particle traps. The crossing section merges five channels into a single stream. In this section the diameter of the different channels vary widely, so that the organic phase containing channels are only 20 µm wide, while the water channels had a width of 100 µm. In addition to these doubled channels the 150 µm wide gas channel is entering the crossing section. The combined stream leaves the crossing section in a straight 2500 µm long and 200 µm wide channel to ensure an optimal mixing process before the liquid leaves the chip.

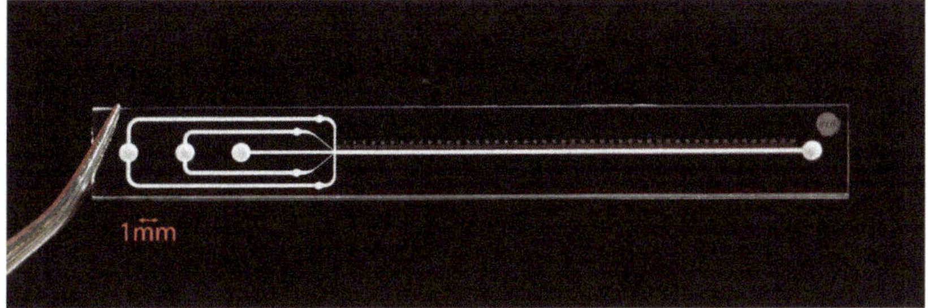

Figure A1. Photograph of the mixer chip

Detection Chip

The critical part of the detection chip is the measurement channel with a nominal depth and width of 2 µm. The channel itself has a length of 400 µm, from which 200 µm are shaped in an U to allow the laser alignment without hitting the supply channels (Figure A2b). The supply channels with a width of 75 µm and a depth of 30 µm allow the regulation and exchange of content of the measurement channel by adjusting the pressure between the four access points. To connect the access points by capillary tubings on the back of the chip, the supply channels spread the access points over the area leading to a length of more than 8 mm for each of them (Figures A2a and A3).

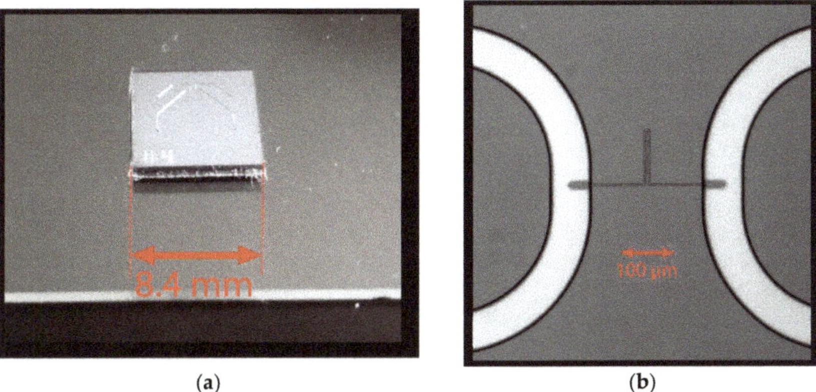

(a)   (b)

Figure A2. (a) Photogrpaph of the detection chip and (b) measurement channel.

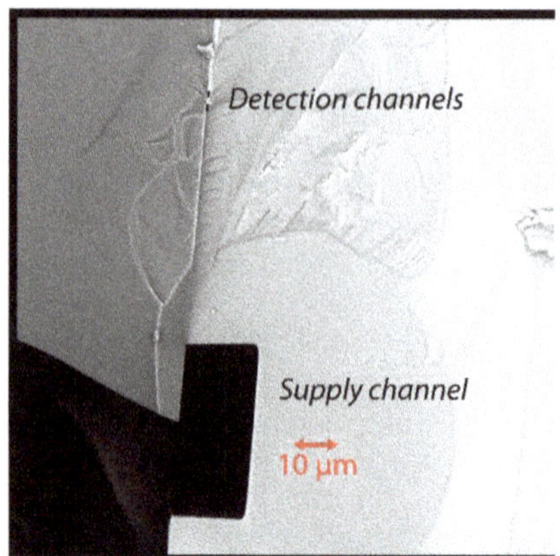

Figure A3. Cross section of channels used in the detection chip.

References

1. Malet-Sanz, L.; Susanne, F. Continuous Flow Synthesis. A Pharma Perspective. *J. Med. Chem.* **2012**, *55*, 4062–4098. [CrossRef] [PubMed]
2. Elvira, K.S.; Solvas, X.C.I.; Wootton, R.C.R.; deMello, A.J. The past, present and potential for microfluidic reactor technology in chemical synthesis. *Nat. Chem.* **2013**, *5*, 905–915. [CrossRef] [PubMed]
3. Plutschack, M.B.; Pieber, B.; Gilmore, K.; Seeberger, P.H. The Hitchhikers Guide in Flow Chemistry. *Chem. Rev.* **2017**, *117*, 11796–11893. [CrossRef] [PubMed]
4. Grundemann, L.; Gonschorowski, V.; Fischer, N.; Scholl, S. Cleaning waste minimization for multiproduct plants: Transferring macro batch to micro conti manufacturing. *J. Clean Prod.* **2012**, *24*, 92–101. [CrossRef]
5. Yoshida, J.; Takahashi, Y.; Nagaki, A. Flash chemistry: Flow chemistry that cannot be done in batch. *Chem. Commun.* **2013**, *49*, 9896–9904. [CrossRef] [PubMed]
6. Krishnadasan, S.; Brown, R.J.C.; deMello, A.J.; deMello, J.C. Intelligent routes to the controlled synthesis of nanoparticles. *Lab Chip* **2007**, *7*, 1434–1441. [CrossRef] [PubMed]
7. Toyota, A.; Nakamura, H.; Ozono, H.; Yamashita, K.; Uehara, M.; Maeda, H. Combinatorial Synthesis of CdSe Nanoparticles Using Microreactors. *J. Phys. Chem. C* **2010**, *114*, 7527–7534. [CrossRef]
8. Sugano, K.; Uchida, Y.; Ichihashi, O.; Yamada, H.; Tsuchiya, T.; Tabata, O. Mixing speed-controlled gold nanoparticle synthesis with pulsed mixing microfluidic system. *Microfluid. Nanofluid.* **2010**, *9*, 1165–1174. [CrossRef]
9. Kunal, P.; Roberts, E.J.; Riche, C.T.; Jarvis, K.; Malmstadt, N.; Bruchtchey, R.L.; Humphrey, S.M. Continuous Flow Synthesis of Rh and RhAg Alloy Nanoparticle Catalysts Enables Scalable Production and Improved Morphological. *Control. Chem. Mat.* **2017**, *20*, 4341–4350. [CrossRef]
10. Duraiswamy, S.; Khan, S.A. Dual-Stage Continuous-Flow Seedless Microfluidic Synthesis of Anisotropic Gold Nanocrystals. *Part. Part. Syst. Charact.* **2014**, *31*, 429–432. [CrossRef]
11. Stolzenburg, P.; Lorenz, T.; Dietzel, A.; Garnweithner, G. Microfluidic synthesis of metal oxide nanoparticles via the nonaqueous method. *Chem. Eng. Sci.* **2018**, *191*, 500–510. [CrossRef]
12. Lorenz, T.; Bojko, S.; Bunjes, H.; Dietzel, A. An inert 3D emulsification device for individual precipitation and concentration of amorphous drug nanoparticles. *Lab Chip* **2018**, *18*, 627–638. [CrossRef] [PubMed]
13. Rhee, M.; Valencia, P.M.; Rodriguez, M.I.; Langer, R.; Farokhzad, O.C.; Karnik, R. Synthesis of Size-Tunable Polymeric Nanoparticles Enabled by 3D Hydrodynamic Flow Focusing in Single-Layer Microchannels. *Adv. Mater.* **2011**, *23*, H79–H83. [CrossRef] [PubMed]
14. Kang, X.; Luo, C.; Wei, Q.; Xiong, C.; Chen, Q.; Chen, Y.; Ouyang, Q. Mass production of highly monodisperse polymeric nanoparticles by parallel flow focusing system. *Microfluid. Nanofluid.* **2013**, *15*, 337–345. [CrossRef]
15. Felnerova, D.; Viret, J.F.; Glück, R.; Moser, C. Liposomes and virosomes as delivery systems for antigens, nucleic acids and drugs. *Curr. Opin. Biotechnol.* **2004**, *15*, 518–529. [CrossRef] [PubMed]

16. Westesen, K.; Bunjes, H.; Koch, M.J.H. Physicochemical characterization of lipid nanoparticles and evaluation of their drug loading capacity and sustained release potential. *J. Control. Release* **1997**, *48*, 223–236. [CrossRef]

17. Mehnert, W.; Mäder, K. Solid lipid nanoparticles Production, characterization and application. *Adv. Drug Deliv. Rev.* **2012**, *47*, 165–196. [CrossRef]

18. Hillaireau, H.; Couvreur, P. Nanocarriers' entry into the cell: Relevance to drug delivery. *Cell. Mol. Life Sci.* **2009**, *66*, 2873–2896. [CrossRef]

19. Zhigaltsev, I.V.; Belliveau, N.; Hafez, I.; Leung, A.K.K.; Huft, J.; Nasen, C.; Cullis, P.R. Bottom-Up Design and Synthesis of Limit Size Lipid Nanoparticle Systems with Aqueous and Triglyceride Cores Using Millisecond Microfluidic Mixing. *Langmuir* **2012**, *28*, 3633–3640. [CrossRef]

20. Lee, J.; Lee, M.G.; Jung, C.; Park, Y.-H.; Song, C.; Coi, M.C.; Park, H.G.; Park, J.K. High-throughput nanoscale lipid vesicle synthesis in a semicircular contraction-expansion array microchannel. *BioChip J.* **2013**, *7*, 210–217. [CrossRef]

21. Xia, H.M.; Sheah, Y.P.; Liu, Y.C.; Wang, W.; Toh, A.G.G.; Wang, Z.P. Anti-solvent precipitation of solid lipid nanoparticles using a microfluidic oscillator mixer. *Microfluid. Nanofluid.* **2015**, *19*, 283–290. [CrossRef]

22. Hood, R.R.; DeVoe, D.L. High-Throughput Continuous Flow Production of Nanoscale Liposomes by Microfluidic Vertical Flow Focusing. *Small* **2015**, *11*, 5790–5799. [CrossRef] [PubMed]

23. Carugo, D.; Bottaro, E.; Owen, J.; Stride, E.; Nastruzzi, C. Liposome production by microfluidics: Potential and limiting factors. *Sci. Rep.* **2016**, *6*, 25876. [CrossRef] [PubMed]

24. Erfle, P.; Riewe, J.; Bunjes, H.; Dietzel, A. Optically monitored segmented flow for controlled ultra-fast mixing and nanoparticle precipitation. *Microfluid. Nanofluid.* **2017**, *21*. [CrossRef]

25. Yurt, A.; Daaboul, G.G.; Connor, J.H.; Goldberg, B.B.; Ünlü, M.S. Single nanoparticle detectors for biological applications. *Nanoscale* **2012**, *4*, 715–726. [CrossRef]

26. Maceiczyk, R.M.; Lignos, I.G.; deMello, A.J. Online detection and automation methods in microfluidic nanomaterial synthesis. *Curr. Opin. Chem. Eng.* **2015**, *8*, 29–35. [CrossRef]

27. Chastek, T.Q.; Iida, K.; Amis, E.J.; Fasolka, M.J.; Beers, K.L. A microfluidic platform for integrated synthesis and dynamic light scattering measurement of block copolymer micelles. *Lab Chip* **2008**, *8*, 950–957. [CrossRef] [PubMed]

28. Destremaut, F.; Salmon, J.-B.; Qi, L.; Chapel, J.P. Microfluidics with on-line dynamic light scattering for size measurements. *Lab Chip* **2009**, *9*, 3289–3296. [CrossRef]

29. Krishnadasan, S.; Tovilla, J.; Vilar, R.; deMello, A.J.; deMello, J.C. On-line analysis of CdSe nanoparticle formation in a continuous flow chip-based microreactor. *J. Mater. Chem.* **2004**, *14*, 2655–2660. [CrossRef]

30. Yue, J.; Falke, F.H.; Schouten, J.C.; Nijhuis, T.A. Microreactors with integrated UV/Vis spectroscopic detection for online process analysis under segmented flow. *Lab Chip* **2013**, *13*, 4855–4863. [CrossRef]

31. Chan, E.M.; Marcus, M.A.; Fakara, S.; EINaggar, M.; Mathies, R.A.; Alivisatos, A.P. Millisecond Kinetics of Nanocrystal Cation Exchange Using Microfluidic X-ray Absorption Spectroscopy. *J. Phys. Chem. A* **2007**, *111*, 12210–12215. [CrossRef] [PubMed]

32. Uehara, M.; Sun, Z.H.; Oyanagi, H.; Yamashita, K.; Fukano, A.; Nakamura, H.; Maeda, H. In situ extended X-ray absorption fine structure study of initial processes in CdSe nanocrystals formation using a microreactor. *Appl. Phys. Lett.* **2009**, *94*. [CrossRef]

33. Huh, D.; Gu, W.; Kamotani, Y.; Grotberg, J.B.; Takayama, S. Microfluidics for flow cytometric analysis of cells and particles. *Physiol. Meas.* **2005**, *26*, R73–R98. [CrossRef] [PubMed]

34. Morgan, H.; Holmes, D.; Green, N.G. High speed simultaneous single particle impedance and fluorescence analysis on a chip. *Curr. Appl. Phys.* **2006**, *6*, 367–370. [CrossRef]

35. Fraikin, J.-L.; Teesalu, T.; McKenney, C.M.; Ruoslathti, E.; Cleland, A.N. A high-throughput label-free nanoparticle analyser. *Nat. Nanotechnol.* **2011**, *6*, 308–313. [CrossRef] [PubMed]

36. Rajauria, S.; Axline, C.; Gottstein, C.; Cleland, N. High-Speed Discrimination and Sorting of Submicron Particles Using a Microfluidic Device. *Nano Lett.* **2015**, *15*, 469–475. [CrossRef] [PubMed]

37. Spencer, D.; Caselli, F.; Bisgena, P.; Morgan, H. High accuracy particle analysis using sheathless microfluidic impedance cytometry. *Lab Chip* **2016**, *16*, 2467–2473. [CrossRef] [PubMed]

38. Carminati, M. Advances in High-Resolution Microscale Impedance Sensors. *J. Sens.* **2017**. [CrossRef]

39. Cui, L.; Zhang, T.; Morgan, H. Optical particle detection integrated in a dielectrophoretic lab-on-a-chip. *J. Micromech. Microeng.* **2002**, *12*, 7–12. [CrossRef]

40. Ignatovich, F.V.; Novotny, L. Real-Time and Background-Free Detection of Nanoscale Particles. *Phys. Rev. Lett.* **2006**, *96*. [CrossRef]

41. Pagliara, S.; Chimerel, C.; Aarts, D.G.A.L.; Langford, R.; Keyser, U.F. Colloid Flow Control in Microchannels and Detection by Laser Scattering. *Prog. Colloid Polym. Sci.* **2012**, *139*, 45–49. [CrossRef]

42. Zhuang, G.; Jensen, T.G.; Kutter, J.P. Detection of unlabeled particles in the low micrometer size range using light scattering and hydrodynamic 3D focusing in a microfluidic system. *Electrophoresis* **2012**, *33*, 1715–1722. [CrossRef] [PubMed]

43. Zhu, S.; Ma, L.; Wang, S.; Chen, C.; Zhang, W.; Yang, L.; Hang, W.; Nolan, J.P.; Wu, L.; Yan, X. Light-Scattering Detection below the Level of Single Fluorescent Molecules for High-Resolution Characterization of Functional Nanoparticles. *ACS Nano* **2014**, *8*, 10998–11006. [CrossRef] [PubMed]

44. Eyal, S.; Quake, S.R. Velocity-independent microfluidic flow cytometry. *Electrophoresis* **2002**, *23*, 2653–2657. [CrossRef]

45. Dittrich, P.S.; Schwille, P. An Integrated Microfluidic System for Reaction, High-Sensitivity Detection, and Sorting of Fluorescent Cells and Particles. *Anal. Chem.* **2003**, *75*, 5767–5774. [CrossRef] [PubMed]

46. Yang, L.; Zhu, S.; Hang, W.; Wu, L.; Yan, X. Development of an Ultrasensitive Dual-Channel Flow Cytometer for the Individual Analysis of Nanosized Particles and Biomolecules. *Anal. Chem.* **2009**, *81*, 2555–2563. [CrossRef] [PubMed]

47. Spencer, D.; Elliott, G.; Morgan, H. A sheath-less combined optical and impedance micro-cytometer. *Lab Chip* **2014**, *14*, 3064–3073. [CrossRef] [PubMed]

48. Friedrich, R.; Block, S.; Alizadehheidari, M.; Heider, S.; Fritzdche, J.; Esbjörner, E.K.; Westerlund, F.; Bally, M. A nano flow cytometer for single lipid vesicle analysis. *Lab Chip* **2017**, *17*, 830–841. [CrossRef] [PubMed]

49. Golden, J.P.; Kim, J.S.; Erickson, J.S.; Hilliard, L.R.; Howell, P.B.; Anderson, G.P.; Nasir, M.; Logler, S. Multi-wavelength microflow cytometer using groove-generated sheath flow. *Lab Chip* **2009**, *9*, 1942–1950. [CrossRef]

50. Simon, P.; Franckowski, M.; Bock, N.; Neukammer, J. Label-free whole blood cell differentiation based on multiple frequency AC impedance and light scattering analysis in a micro flow cytometer. *Lab Chip* **2016**, *16*, 2326–2338. [CrossRef]

51. Mitra, A.; Ignatovich, F.; Novotny, L. Real-time optical detection of single human and bacterial viruses based on dark-field interferometry. *Biosens. Bioelectron.* **2012**, *31*, 499–504. [CrossRef] [PubMed]

52. Haiden, C.; Wopelka, T.; Jechm, M.; Puchberger-Enengl, D.; Weber, E.; Keplinger, F.; Vellenloop, M.J. A microfluidic system for visualisation of individual sub-micron particles by light scattering. *Procedia Eng.* **2012**, *47*, 680–683. [CrossRef]

53. D'Addio, S.M.; Prud'homme, R.K. Controlling drug nanoparticle formation by rapid precipitation. *Adv. Drug Deliv. Rev.* **2011**, *63*, 417–426. [CrossRef] [PubMed]

54. Pagliara, S.; Chimerel, C.; Langford, R.; Aarts, D.; Keyser, U.F. Parallel sub-micrometre channels with different dimensions for laser scattering detection. *Lab Chip* **2011**, *11*, 3365–3368. [CrossRef] [PubMed]

55. Greenspan, P.; Fowler, S.D. Spectrofluorometric studies of the lipid probe, nile red. *J. Lipid Res.* **1985**, *26*, 781–789.

56. Al-Halhouli, A.; Al-Faqheri, W.; Alhamarneh, B.; Hecht, L.; Dietzel, A. Spiral Microchannels with Trapezoidal Cross Section Fabricated by Femtosecond Laser Ablation in Glass for the Inertial Separation of Microparticles. *Micromachines* **2018**, *9*, 171. [CrossRef] [PubMed]

57. Schulze, T.; Mattern, K.; Fruh, E.; Hecht, L.; Rustenbeck, I.; Dietzel, A. A 3D microfluidic perfusion system made from glass for multiparametric analysis of stimulus-secretioncoupling in pancreatic islets. *Biomed. Microdevices* **2017**, *19*, 47. [CrossRef] [PubMed]

Article

Affordable Fabrication of Conductive Electrodes and Dielectric Films for a Paper-Based Digital Microfluidic Chip

Veasna Soum [1], Yunpyo Kim [1], Sooyong Park [1], Mary Chuong [1], Soo Ryeon Ryu [1], Sang Ho Lee [2], Georgi Tanev [3], Jan Madsen [3], Oh-Sun Kwon [1,*] and Kwanwoo Shin [1,*]

[1] Department of Chemistry, Institute of Biological Interfaces, Sogang University, Seoul 04107, Korea; veasna_soum@yahoo.com (V.S.); hhh0411hhh@naver.com (Y.K.); qkrtndyd111@naver.com (S.P.); marychuong05@gmail.com (M.C.); navio23@sogang.ac.kr (S.R.R.)

[2] Department of Chemical Engineering, The Cooper Union for Advancement of Science and Art, New York, NY 10003, USA; private.shlee@gmail.com

[3] Department of Applied Mathematics and Computer Science, Technical University of Denmark, 2800 Lyngby, Denmark; g.p.tanev@gmail.com (G.T.); jama@dtu.dk (J.M.)

* Correspondence: oskwon@sogang.ac.kr (O.-S.K.); kwshin@sogang.ac.kr (K.S.)

Received: 3 January 2019; Accepted: 3 February 2019; Published: 7 February 2019

Abstract: In order to fabricate a digital microfluidic (DMF) chip, which requires a patterned array of electrodes coated with a dielectric film, we explored two simple methods: Ballpoint pen printing to generate the electrodes, and wrapping of a dielectric plastic film to coat the electrodes. For precise and programmable printing of the patterned electrodes, we used a digital plotter with a ballpoint pen filled with a silver nanoparticle (AgNP) ink. Instead of using conventional material deposition methods, such as chemical vapor deposition, printing, and spin coating, for fabricating the thin dielectric layer, we used a simple method in which we prepared a thin dielectric layer using pre-made linear, low-density polyethylene (LLDPE) plastic (17-μm thick) by simple wrapping. We then sealed it tightly with thin silicone oil layers so that it could be used as a DMF chip. Such a treated dielectric layer showed good electrowetting performance for a sessile drop without contact angle hysteresis under an applied voltage of less than 170 V. By using this straightforward fabrication method, we quickly and affordably fabricated a paper-based DMF chip and demonstrated the digital electrofluidic actuation and manipulation of drops.

Keywords: dielectric film; plastic wrap; ballpoint pen printing; conductive electrode; digital microfluidic chip; electrowetting

---

1. Introduction

A few decades ago, a digital microfluidic (DMF) chip was introduced as a new type of lab-on-a-chip (LOC) device [1–3]. The DMF chip manipulates droplets on the surface of a set of electrode arrays coated with a dielectric film and actuated by applying an electrical potential. The actuation principle is based on the so-called electrowetting on dielectric (EWOD) phenomena. EWOD is a very practical way for fluidic manipulation in microfluidic devices. It has been used for creating an electrowetting valve to control the flow of the fluid in a continuous-flow paper-based microfluidic device [4,5]. Moreover, EWOD has been effectively used for the mixing, splitting, and transporting of aqueous samples, which are essential characteristics for LOCs [6–8]. That type of LOC was a simplified and minimized device, because many scalable components, such as complex pumps, guiding channels and valves, had been removed and replaced with planar structures capable of actuating a DMF drop driven by the EWOD phenomena [9–11].

For the fabrication of a planar DMF chip, a set of patterned electrode arrays and a thin dielectric layer must be fabricated [1–3]. Many deposition methods can be used to pattern electrodes, but most rely on the phases of the deposited materials, such as metals with lithography and sputtering, wet-based inks with inkjet printing, copper-based printed circuit boards with etching, and so on [9,12,13]. Particularly, a paper-based DMF chip, which is much easier to implement for inkjet printing of patterned electrodes than any other conventional substrates, such as wafers, glasses and polychlorinated biphenyls (PCBs), was introduced, providing an easy and convenient means of fabricating DMF LOCs [8,14–17]. Commonly, thin dielectric layers for DMF chips are generated by using spin coating and chemical vapor deposition (CVD), which require a clean room facility [18,19]. These processes cause the fabrication of DMF chips to be expensive, even though the cost of LOC devices for point-of-care (POC) applications should be inexpensive.

Here we introduce and exploit a simplified way of printing that uses a ballpoint pen filled with ink made of a conductive material, as well as a digital plotter for printing electrode arrays on paper for a DMF chip. Moreover, even though parylene-C has been widely used for the deposition of dielectric films, especially when the CVD method is used, we attempted to devise a significantly simpler way to fabricate dielectric films; to that end, we explored a wrapping method that uses a pre-made plastic film. By combining the above two simplified deposition methods to deposit a double-layer, i.e., a dielectric layer top-coated onto a conductive layer, we affordably fabricated a paper-based DMF chip and successfully demonstrated its good electrofluidic performance and operation.

2. Materials and Methods

2.1. Chemicals and Materials

Silver nanoparticle (AgNP) ink was purchased from Advance Nano Products Co., Ltd (DGP 40LT-15C, Sejong, Korea). The surface tension and the average particle size were 22 mN/m and less than 50 nm, respectively. For the ink cartridge and the housing barrel of the ballpoint pen, we used the cartridge from a ball pen (ball diameter of 1.0 mm, Zebra, Tokyo, Japan) and the barrel from a permanent marker (Monami, Yongin-si, Korea), respectively. Inkjet photo paper (C13S042187, Epson, Tokyo, Japan) was selected for the printing substrate because it is glossy and thermally stable up to a temperature of 200 °C. A digital plotter (Cricut Explore Air, Provo Craft & Novelty, Inc., South Jordan, UT, USA) with software (Cricut Design Space, Ver. 3) was employed for the printing. The ballpoint pen used for the printing with a digital plotter was prepared as reported previously [20]. After the original ink had been removed, the empty cartridge was sonicated in an ethanol bath for a day, after which it was cleaned with water. The cleaned cartridge was then filled by pipetting with 80 μL of AgNP ink. A plastic film, which is used to wrap food in a home kitchen (thickness of ~17 μm, Cleanwrap, Seoul, Korea) and is made of flexible, stretchable linear low-density polyethylene (LLDPE) was used. For both the lubricant and the adhesive filler, dielectric silicone oil with a dynamic viscosity of 10 cP (Sigma-Aldrich, St. Louis, MO, USA) was used.

2.2. Printing of Electrode Arrays for Paper-Based DMF Chip

In the patterning of electrodes for the paper-based DMF chip, we used AgNP ink, which is a highly conductive material, a ballpoint pen, and a digital plotter. The electrode arrays for the paper-based DMF chips were designed using computer-aided design software (Adobe Illustrator CC 2015, Adobe, San Jose, CA, USA) and then exported to an AutoCAD Interchange file (*.DXF). Before the printing, the prepared ballpoint pen was inserted into clamp A of the digital plotter. Then the design file was uploaded to an online design space of the digital plotter and set as the writing function. We set the printing file as the writing function (Clamp A) with a printing speed of 5 cm/s. The printing speed was fixed at 4.6 cm/s for a horizontal or vertical line and at 5.9 cm/s for a 45°-tilted line. After the printing, the AgNP electrode arrays were annealed at 170 °C for 30 min to reduce electrical resistance.

We characterized the printed patterns by using a field emission scanning electron microscope (FE-SEM) (JSM-7100F, JEOL, Pleasanton, CA, USA).

2.3. Preparation of Thin Dielectric Film for Paper-Based DMF Chip

The dielectric layer for the paper-based DMF chip was prepared using a commercial pre-made plastic wrap (17 μm) made of LLDPE. It was prepared by fixing it to an adhesive plastic frame to make it flat. We applied silicone oil (20 μL) to cover the entire surface of the LLDP-dielectric layer to reduce the layer's surface friction during the movement of a droplet on the surface. We applied a thin layer of silicone oil to the printed electrodes and to the substrate to ensure that the LLDPE-dielectric layer bonded to the printed electrodes and the substrate with no air bubbles at the interface. The surface properties of the LLDPE-dielectric film were observed by using atomic force microscopy (AFM) (NanoStation, Pucotech, Seoul, Korea) and Fourier-transform infrared spectroscopy (FT-IR) (Agilent Technologies, Cray 640 FTIR, Santa Clara, CA, USA). The contact angles (CA) of droplets were measured by using a freeware program (ImageJ, 1.51p, National Institutes of Health (NIH), Bethesda, MD, USA) on images of water droplets captured by using a contact angle analyzer (Phx 300, Image XP 5.6U, SEO, Seoul, Korea). The leakage of electrical current was measured by using a source meter (Keithley, 2400, Keithley Instruments, Solon, OH, USA).

3. Results

3.1. Electrode Arrays of the Paper-Based DMF Chip

In the fabrication of the DMF chip, the electrode size and shape must be carefully designed and depend on the sample volume and whether the system is closed or open. Obtaining a proper method for printing a customized electrode is crucial. To that end, we explored a simple printing method using a customized ballpoint pen [20–22] and a type of direct contact printing in which we simply replaced the pigment ink in the stylus with conductive AgNP ink. After the ballpoint pen had been prepared, we inserted it into a pen holder of the plotter for printing in a programmable manner [20]. The contact pressure and the lateral movement of the ballpoint pen controlled by the digital plotter allowed the deposition gap of the ballpoint pen to open, and rotated the ballpoint of the ballpoint pen. As a result, AgNP ink was deposited on the printing substrate. Figure 1 shows the printing setup for patterning electrode arrays for a paper-based DMF by using a ballpoint pen and a plotter. With the printing system, a layman can print a customized design on paper in minutes without the need for expensive equipment (Video S1).

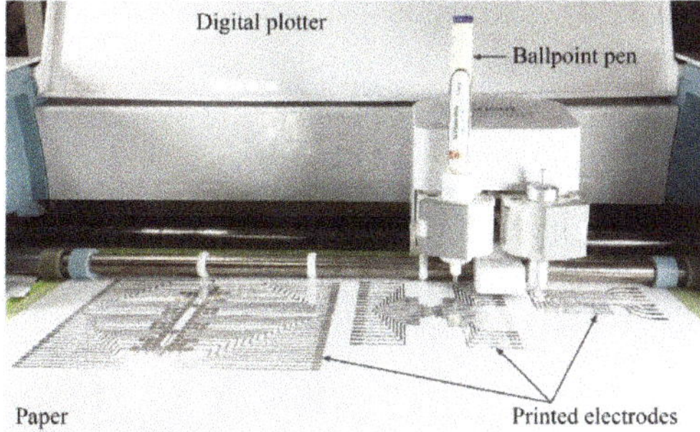

Figure 1. Printing setup for patterning electrode arrays on paper for the fabrication of a digital microfluidic (DMF) chip by using a ballpoint pen and a digital plotter.

In Figure 2a,b, we demonstrate the printing on paper of AgNP electrodes with the various sizes and shapes that are commonly used in DMF chips and with the desired designs. However, if a large electrode is to be printed, the printed electrode has to be designed as a group of lines [20]. The printed electrodes showed clean edges and good resolution, and by printing the AgNP electrodes using a 1-mm ballpoint-pen diameter, we were able to generate in a single printing an electrode with a minimum width and gap of 850 ± 50 µm and 200 ± 50 µm, respectively (Figure 2c). The SEM image shows that the printed electrode contained a high density of AgNPs connected to their neighbors (Figure 2d). Moreover, the thickness of the printed electrode (single printing) was 450 ± 50 nm, which is suitable for DMF use (Figure 2e).

Because the printed AgNPs are, manifestly, neither strongly metallic bonded nor perfectly crystallized, but rather a grouped domain of aggregated nanoparticle grains, in order to enhance the conductivity of the AgNP film and to make it more tightly bonded, we thermally annealed the printed AgNPs at 170 °C for 30 min. This low-temperature annealing was selected in consideration of the limited thermal resistance of the photo paper (210 °C) (Figure 2c). With our setup, the electrical resistance of the AgNP pattern was about 16 Ω/cm for one printing, was reduced to about 7 Ω/cm for two printings and was slightly less for three and four printings (Figure 2f). The slight reduction in electrical resistance for three and four printings is because the first printed layer can be squeezed by a ballpoint pen during the printing of a next printed layer. This is a limitation of the contact printing technique in the printing of multi-layer patterns. Although increasing the number of printings increased the electrical conductivity of the printed pattern, with our printing setup, one printing was sufficient to generate a highly conductive electrode for a DMF chip. These results clearly show that our printing method has the ability to fabricate electrodes in arbitrary patterns. Furthermore, the deposition of a thin electrode is interesting because such an electrode has definite advantages in many applications, including DMF chips.

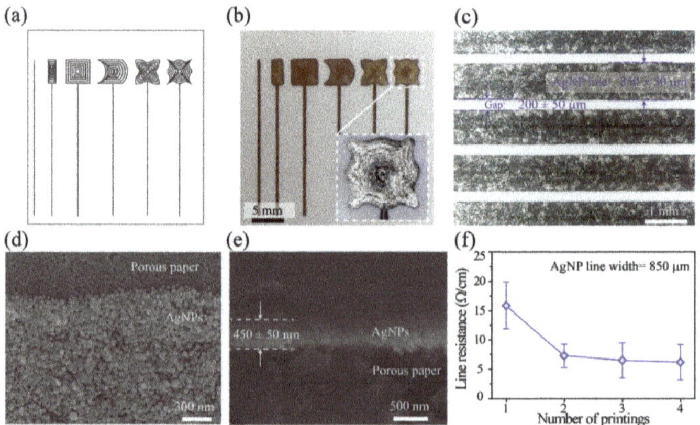

Figure 2. Printing silver nanoparticle (AgNP) electrodes on paper. (**a**) Electrode designs with various sizes and shapes, (**b**) electrodes printed with the designs in (**a**) and (inset) an enlargement of one design, and (**c**) printed AgNP lines. SEM images of the printed pattern: (**d**) top view and (**e**) cross-sectional view. (**f**) Surface electrical resistance of the printed AgNP line versus the number of printings.

3.2. Dielectric Layer of the Paper-Based DMF Chip

Our easy printing of conductive materials to obtain precisely patterned electrodes meets the challenge of being a simple coating process for achieving highly hydrophobic dielectric films and offers an affordable way to fabricate paper-based DMF chips. The intervening dielectric thin film allows the use of high voltages for inducing a stronger electrowetting force so that the shape of the drop can be sufficiently deformed, and the drop can eventually move [3]. However, since Berge introduced an excellent dielectric material, parylene-C, which has a very high electrical breakdown voltage (200 V/µm), new attempts to obtain alternative dielectrics have been surprisingly rare, except for a few materials, such as Teflon, SU-8, CYTOP, and polydimethylsiloxane (PDMS) [3,19,23–26]. Even worse, most thin-film deposition methods, such as CVD for parylene-C and spin-coating for Teflon, require heavy instruments, including gas- and temperature-control systems on a lab-scale, which is a severe obstacle to the affordable fabrication of DMF chips.

Because of the above factors, we explored a new approach, the hand-wrapping approach, to fabricate a dielectric film for a paper-based DMF chip by using a commercially available plastic film made of LLDPE. Originally, this method was developed by Gaudi Labs in 2017 and was then made available to the public [24]. Even though commercial, pre-made LLDPE wrap is low-cost, the pristine wrap cannot be used for the dielectric layer of a DMF chip without treatment. AFM data showed that the surface of the pre-made LLDPE film contained some defects (Figure 3a,b). These defects may allow electrical current leakage during the operation of a DMF chip under high voltage. Low hydrophobicity (contact angle: ~93°) is another drawback of the pre-made LLDPE film; this causes the surface friction to increase, thereby impeding the movement of drops during DMF chip operation. To overcome these limitations, we applied a thin layer of silicone oil on the LLDPE-dielectric film in order to make the surface slippery, thus reducing the surface friction [27–29] (Figure 3c). For ease of handling and the avoidance of wrinkles, we used an adhesive plastic frame, to which the LLDPE-dielectric film was fixed to make it flat (Figure 3c). With the FT-IR spectrum, we confirmed the presence of a thin layer of silicone oil on the LLDPE-dielectric film (Figure 4d). We believe that covering the LLDPE film with a thin layer of silicone oil seals the surface defects on the surface of pre-made LLDPE, which should minimize the leakage of electrical current during high-voltage DMF chip operation. We observed the leakage current across the LLDPE-dielectric layer treated with silicone oil by applying various voltages from 50 V to 200 V (Figure 4a). The leakage current density dramatically increased when 50 V to 175 V were applied, and rapidly increased when a higher voltage was applied. However, this low leakage of electrical current can be ignored if the operation voltage is lower than 200 V. With our method, we were able to rapidly prepare and apply a dielectric layer for our fabricated paper-based DMF chip without the need for any special tools or devices.

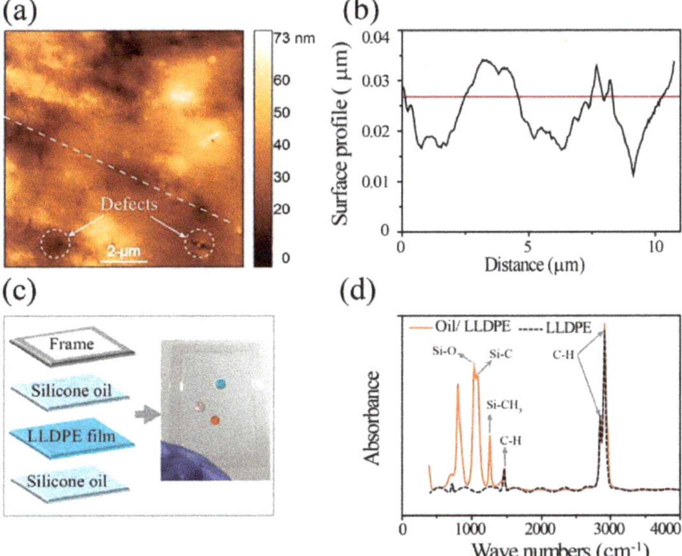

Figure 3. Surface morphology of the commercial LLDPE film. (**a**) Atomic force microscopy (AFM) image of the film and (**b**) representative surface profile extracted from (a) at the dashed line. (**c**) Schematic structure of a dielectric film prepared for the paper-based DMF chip and a photograph of the prepared dielectric film with three colorful drops on it. (**d**) FT-IR spectrum of a LLDPE-dielectric film with and without a silicone oil coated.

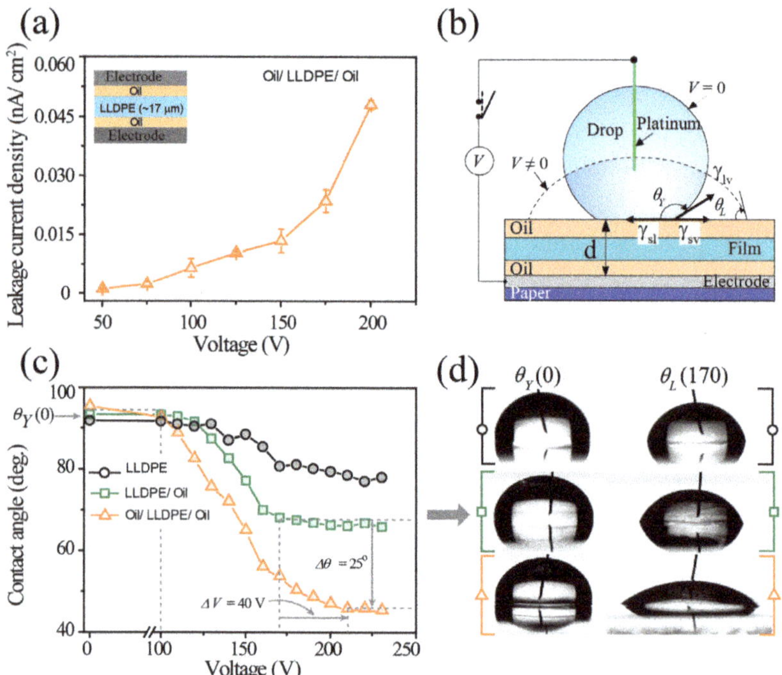

Figure 4. (**a**) Leakage current density across the LLDPE-dielectric film and its schematic measurement (inset). Characterization of the electrowetting on dielectric (EWOD) experiment on the LLDPE-dielectric film: (**b**) EWOD setup, and (**c,d**) the electro-spreading of the drop under applied voltages.

3.3. Droplet Actuation on the Paper-Based DMF Chip

For a sessile drop on the LLDPE-wrapped electrodes, we measured the CA as a function of the applied voltage (Figure 4b) so as to investigate the dielectric properties of the LLDPE-dielectric film. As shown in Figure 4c, the initial CA at no voltage was greater than 90°, indicating that the hydrophobicity of the film was sufficient for EWOD applications (black, Figure 4c). However, the CA response to changes in the applied voltage was irregular, not smooth, the so-called CA hysteresis [3], mostly due to the existence of an air gap under the LLDPE-dielectric film. In order to remove the air gap, we coated the bottom side of the LLDPE-dielectric film with a dielectric liquid sealant, silicone oil, to a thickness of roughly 200 nm [3,14,24]. After this sealant treatment, we obtained a much-improved CA response (green in Figure 4c), and the variation in the CA with applied voltage was small, being only $\Delta\theta = 93 - 68 = 25°$ for $\Delta V = 100 - 170 = -70$ Vdc. Of course, we ignored both the initial inert wetting region at low voltages < 100 Vdc (leftmost dashed line in Figure 4c), and the saturated CA region at high voltages > 170 Vdc (middle dashed line in Figure 4c). To increase the differential CA, $\Delta\theta(V)$, we also coated the top of the wrapped film with silicone oil to lubricate the surface. As a result, $\Delta\theta(V)$ was doubled from 25° to 50° for $\Delta V = -40$ Vdc (orange in Figure 4c). Figure 4d shows representative images of sessile drops wetted on the surface of the LLDPE-dielectric films at zero and 170 V taken from Figure 4c.

According to the Young-Lippmann equation describing the relationship between the Young and the Lippmann contact angles, $\theta_Y(0)$ and $\theta_L(V)$, for the equilibrium states of the drop,

$$\cos\theta_L(V) = \cos\theta_Y(0) + \frac{1}{2\gamma_{lv}}CV^2 \tag{1}$$

where γ_{lv} is the interfacial tension between the liquid and the vapor. A higher voltage is needed to compensate for the excessive thickness of the plastic wrapped film, 17 µm, because of the capacitance $C = \varepsilon A/d$, where A, d and ε are the area, thickness, and permittivity of the capacitor [3,23,30]. Fortunately, however, due to the quadratic dependence on the voltage, the required voltage for our excessively thickly wrapped film was still low in the range of less than 210 Vdc (rightmost dashed line in Figure 4c), which can be provided by a small-size electronic power supply. If this were not the case, electric breakdown might occur because the required voltage for a 17-µm thickness would be 425 V due to the relatively low dielectric strength of 25 V/µm of the LLDPE-dielectric film.

3.4. Operation of the Paper-Based DMF Chip

Finally, using the same multilayer configuration for the EWOD experiment (Figure 4b), except for replacing the single electrode with a trail-patterned array of electrodes, we fabricated a paper-based DMF chip easily, rapidly and affordably. The paper-based DMF chip (area 5 cm × 5 cm) demonstrated in Figure 5a costs less than USD 1 to make with a setup that costs approximately USD 241. Because a high voltage ($V_{dc} > 170$ V) caused the drop's spread to reach sub-maximal CA saturation, we operated our paper-based DMF chip at a voltage of 170 V. With the applied voltage, the electrode arrays (width: 800 µm, gap: 400 µm) could move a 5-µL drop along the activated electrode (yellow mark, inset of Figure 5a). The drop was sufficiently deformed and touched the adjacent electrode, eventually moving toward the activated electrode. The drop's movement across the planar electrodes can be attributed to the large differential variation of CA, $\Delta\theta = 50\,°C$ (orange, Figure 4c) because the actuation EWOD force is proportional to $\Delta\theta$ according to Furmidge's equation [30,31]. Similarly, we successfully demonstrated digital fluidic manipulation along the trail-patterned electrodes.

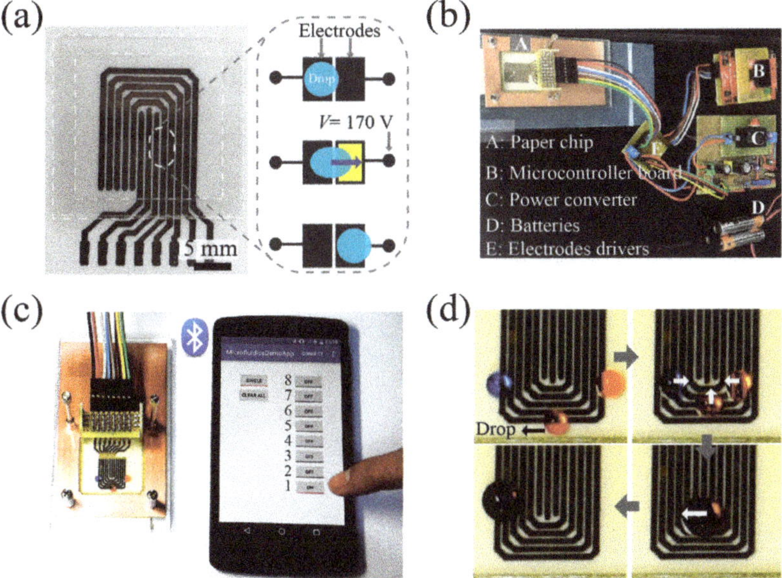

Figure 5. Demonstration of drop actuation on an affordable paper-DMF chip: (**a**) a printed paper chip and (inset) drop actuation scheme and (**b**) the chip connected to a power switching device. (**c**) A smartphone App being run through wireless Bluetooth, and (**d**) three digital drops initially at rest being simultaneously transported (top) to merge into one drop at the center and then being transported to the left (bottom).

A 15-µL drop was formed by transporting and merging three 5-µL drops along different paths after the paper-based DMF chip had been connected to a custom designed control system (Figure 5d, Video S2). The control system consists of a hardware prototype and a smartphone, as shown in Figure 5b,c. The smartphone runs a custom-developed app that can send commands over a Bluetooth link to the hardware prototype. The hardware prototype is based on the Stelaris LM4F120 Launchpad (Texas Instruments, Dallas, TX, USA) development board, which is interfaced with a Bluetooth module, a display, a custom-developed isolated flyback boost converter, and an electrode driver board based on a commercially available integrated circuit (HV2201, Microchip, Chandler, AZ, USA). The control system design is similar to that in our previous reports [8,32]. Our paper-based DMF chip can be used more than 100 times without damage. In addition, it is a benefit of the design that the dielectric layer can be substituted, which is different from the parylene coatings that are attached to the surface.

4. Conclusions

In conclusion, two affordable methods for generating both conductive electrode arrays and dielectric layers for use in the fabrication of paper-based DMF chips were introduced. The conductive electrode arrays for the DMF chips were printed according to a program on paper by using the prepared ballpoint pen and a digital plotter. In the preparation of the dielectric layer for the paper-based DMF chip, we used commercial food wrap made of LLDPE. We framed and treated the LLDPE-dielectric film with a thin layer of silicone oil on both surfaces (front and back). The treatment provided a slippery surface and covered the surface defects in the LLDPE-dielectric film, thereby preventing the leakage of current during chip operation. We investigated the properties of the LLDPE-dielectric film, and the results showed that the film sufficiently satisfied the dielectric requirements for a multilayer DMF chip. This approach holds much promise as a simple, easy, and rapid way to fabricate paper-based DMF chips affordably.

Supplementary Materials: The following are available online at http://www.mdpi.com/2072-666X/10/2/109/s1, Video S1: Printing on paper the conductive electrode array for the DMP, Video S2: Operation of the DMF with a smartphone App.

Author Contributions: Conceptualization: V.S.; data curation: Y.K., S.P., M.C. and S.H.L.; formal analysis: V.S., S.R.R. and O.-S.K.; methodology: V.S. and G.T.; resources: K.S.; software: G.T. and J.M.; supervision: K.S.; validation: K.S.; visualization: Y.K., S.P., M.C. and S.H.L.; writing the original draft: O.-S.K.; writing, reviewing and editing the final manuscript: V.S., J.M., O.-S.K. and K.S.

Funding: The authors gratefully acknowledge support from the Basic Science Research Program (2017R1D1A1B03032095 and 2016R1A2B3015239) through the National Research Foundation of Korea (NRF) funded by the Ministry of Science and ICT (MSIT).

Acknowledgments: We thank Rüdiger Trojok in Hackteria for technical comments.

Conflicts of Interest: The authors declare no conflicts of interest.

References

1. Pollack, M.G.; Fair, R.B.; Shenderov, A.D. Electrowetting-based actuation of liquid droplets for microfluidic applications. *Appl. Phys. Lett.* **2000**, *77*, 1725–1726. [CrossRef]
2. Moon, H.; Cho, S.K.; Garrell, R.L.; Kim, C.J. Low voltage electrowetting-on-dielectric. *J. Appl. Phys.* **2002**, *92*, 4080–4087. [CrossRef]
3. Frieder, M.; Jean-Christophe, B. Electrowetting: From basics to applications. *J. Phys. Condens. Matter* **2005**, *17*, R705.
4. Ainla, A.; Hamedi, M.M.; Guder, F.; Whitesides, G.M. Electrical Textile Valves for Paper Microfluidics. *Adv. Mater.* **2017**, *29*, 1702894. [CrossRef] [PubMed]
5. Koo, C.K.; He, F.; Nugen, S.R. An inkjet-printed electrowetting valve for paper-fluidic sensors. *Analyst* **2013**, *138*, 4998–5004. [CrossRef] [PubMed]
6. Mugele, F.; Staicu, A.; Bakker, R.; van den Ende, D. Capillary Stokes drift: A new driving mechanism for mixing in AC-electrowetting. *Lab Chip* **2011**, *11*, 2011–2016. [CrossRef] [PubMed]

7.   Geng, H.; Feng, J.; Stabryla, L.M.; Cho, S.K. Dielectrowetting manipulation for digital microfluidics: Creating, transporting, splitting, and merging of droplets. *Lab Chip* **2017**, *17*, 1060–1068. [CrossRef]

8.   Ruecha, N.; Lee, J.; Chae, H.; Cheong, H.; Soum, V.; Preechakasedkit, P.; Chailapakul, O.; Tanev, G.; Madsen, J.; Rodthongkum, N.; et al. Paper-Based Digital Microfluidic Chip for Multiple Electrochemical Assay Operated by a Wireless Portable Control System. *Adv. Mater. Technol.* **2017**, *2*, 1600267. [CrossRef]

9.   Samiei, E.; Tabrizian, M.; Hoorfar, M. A review of digital microfluidics as portable platforms for lab-on a-chip applications. *Lab Chip* **2016**, *16*, 2376–2396. [CrossRef]

10.   Wang, H.; Chen, L.; Sun, L. Digital microfluidics: A promising technique for biochemical applications. *Front. Mech. Eng.* **2017**, *12*, 510–525. [CrossRef]

11.   Malic, L.; Brassard, D.; Veres, T.; Tabrizian, M. Integration and detection of biochemical assays in digital microfluidic LOC devices. *Lab Chip* **2010**, *10*, 418–431. [CrossRef] [PubMed]

12.   Ehsan, S.; Mina, H. Systematic analysis of geometrical based unequal droplet splitting in digital microfluidics. *J. Micromech. Microeng.* **2015**, *25*, 055008.

13.   Sista, R.S.; Wang, T.; Wu, N.; Graham, C.; Eckhardt, A.; Winger, T.; Srinivasan, V.; Bali, D.; Millington, D.S.; Pamula, V.K. Multiplex newborn screening for Pompe, Fabry, Hunter, Gaucher, and Hurler diseases using a digital microfluidic platform. *Clin. Chim. Acta* **2013**, *424*, 12–18. [CrossRef] [PubMed]

14.   Ko, H.; Lee, J.; Kim, Y.; Lee, B.; Jung, C.H.; Choi, J.H.; Kwon, O.S.; Shin, K. Active digital microfluidic paper chips with inkjet-printed patterned electrodes. *Adv. Mater.* **2014**, *26*, 2335–2340. [CrossRef] [PubMed]

15.   Fobel, R.; Kirby, A.E.; Ng, A.H.; Farnood, R.R.; Wheeler, A.R. Paper microfluidics goes digital. *Adv. Mater.* **2014**, *26*, 2838–2843. [CrossRef] [PubMed]

16.   Jang, I.; Ko, H.; You, G.; Lee, H.; Paek, S.; Chae, H.; Lee, J.H.; Choi, S.; Kwon, O.S.; Shin, K.; et al. Application of Paper EWOD (Electrowetting-on-Dielectrics) Chip: Protein Tryptic Digestion and its Detection Using MALDI-TOF Mass Spectrometry. *Biochip J.* **2017**, *11*, 146–152. [CrossRef]

17.   Jafry, A.T.; Lee, H.; Tenggara, A.P.; Lim, H.; Moon, Y.; Kim, S.H.; Lee, Y.; Kim, S.M.; Park, S.; Byun, D.; et al. Double-sided electrohydrodynamic jet printing of two-dimensional electrode array in paper-based digital microfluidics. *Sens. Actuators B-Chem.* **2019**, *282*, 831–837. [CrossRef]

18.   Ng, A.H.; Choi, K.; Luoma, R.P.; Robinson, J.M.; Wheeler, A.R. Digital microfluidic magnetic separation for particle-based immunoassays. *Anal. Chem.* **2012**, *84*, 8805–8812. [CrossRef]

19.   Sathyanarayanan, G.; Haapala, M.; Sikanen, T. Interfacing Digital Microfluidics with Ambient Mass Spectrometry Using SU-8 as Dielectric Layer. *Micromachines* **2018**, *9*, 649. [CrossRef]

20.   Soum, V.; Cheong, H.; Kim, K.; Kim, Y.; Chuong, M.; Ryu, S.R.; Yuen, P.K.; Kwon, O.-S.; Shin, K. Programmable Contact Printing Using Ballpoint Pens with a Digital Plotter for Patterning Electrodes on Paper. *ACS Omega* **2018**, *3*, 16866–16873. [CrossRef]

21.   Tai, Y.L.; Yang, Z.G. Fabrication of paper-based conductive patterns for flexible electronics by direct-writing. *J. Mater. Chem.* **2011**, *21*, 5938–5943. [CrossRef]

22.   Russo, A.; Ahn, B.Y.; Adams, J.J.; Duoss, E.B.; Bernhard, J.T.; Lewis, J.A. Pen-on-paper flexible electronics. *Adv. Mater.* **2011**, *23*, 3426–3430. [CrossRef] [PubMed]

23.   Fobel, R.; Fobel, C.; Wheeler, A.R. DropBot: An open-source digital microfluidic control system with precise control of electrostatic driving force and instantaneous drop velocity measurement. *Appl. Phys. Lett.* **2013**, *102*, 193513. [CrossRef]

24.   Alistar, M.; Gaudenz, U. OpenDrop: An Integrated Do-It-Yourself Platform for Personal Use of Biochips. *Bioengineering* **2017**, *4*, 45. [CrossRef]

25.   Dixon, C.; Ng, A.H.; Fobel, R.; Miltenburg, M.B.; Wheeler, A.R. An inkjet printed, roll-coated digital microfluidic device for inexpensive, miniaturized diagnostic assays. *Lab Chip* **2016**, *16*, 4560–4568. [CrossRef] [PubMed]

26.   Abadian, A.; Sepehri Manesh, S.; Jafarabadi Ashtiani, S. Hybrid paper-based microfluidics: Combination of paper-based analytical device (μPAD) and digital microfluidics (DMF) on a single substrate. *Microfluid. Nanofluid.* **2017**, *21*, 65. [CrossRef]

27.   Cheong, H.; Oh, H.; Kim, Y.; Kim, Y.; Soum, V.; Choi, J.H.; Kwon, O.S.; Shin, K. Effects of Silicone Oil on Electrowetting to Actuate a Digital Microfluidic Drop on Paper. *J. Nanosci. Nanotechnol.* **2018**, *18*, 7147–7150. [CrossRef]

28.   Liu, Y.G.; Banerjee, A.; Papautsky, I. Precise droplet volume measurement and electrode-based volume metering in digital microfluidics. *Microfluid. Nanofluid.* **2014**, *17*, 295–303. [CrossRef]

29. Banerjee, A.; Noh, J.H.; Liu, Y.G.; Rack, P.D.; Papautsky, I. Programmable Electrowetting with Channels and Droplets. *Micromachines* **2015**, *6*, 172–185. [CrossRef]
30. Ko, H.; Lee, J.S.; Jung, C.H.; Choi, J.H.; Kwon, O.S.; Shin, K. Actuation of digital micro drops by electrowetting on open microfluidic chips fabricated in photolithography. *J. Nanosci. Nanotechnol.* **2014**, *14*, 5894–5897. [CrossRef]
31. Biole, D.; Bertola, V. The role of the microscale contact line dynamics in the wetting behaviour of complex fluids. *Arch. Mech.* **2015**, *67*, 401–414.
32. Tanev, G.; Madsen, J. A correct-by-construction design and programming approach for open paper-based digital microfluidics. In Proceedings of the 2017 Symposium on Design, Test, Integration and Packaging of MEMS/MOEMS (DTIP), Bordeaux, France, 29 May–1 June 2017; pp. 1–6.

Article

Magnetically Induced Flow Focusing of Non-Magnetic Microparticles in Ferrofluids under Inclined Magnetic Fields

Laan Luo and Yongqing He *

School of Chemical Engineering, Kunming University of Science and Technology, Kunming 650500, China; lla1016@163.com
* Correspondence: yqhe@kmust.edu.cn

Received: 30 December 2018; Accepted: 14 January 2019; Published: 15 January 2019

Abstract: The ability to focus biological particles into a designated position of a microchannel is vital for various biological applications. This paper reports particle focusing under vertical and inclined magnetic fields. We analyzed the effect of the angle of rotation (θ) of the permanent magnets and the critical Reynolds number (Re_c) on the particle focusing in depth. We found that a rotation angle of 10° is preferred; a particle loop has formed when $Re < Re_c$ and Re_c of the inclined magnetic field is larger than that of the vertical magnetic field. We also conducted experiments with polystyrene particles (10.4 μm in diameter) to prove the calculations. Experimental results show that the focusing effectiveness improved with increasing applied magnetic field strength or decreasing inlet flow rate.

Keywords: flow focusing; magnetic field; microparticles; ferrofluids

---

1. Introduction

Continuous flow focusing of microparticles/cells is an essential step for the downstream counting [1] and analyses [2] in microfluidics. Since it allows focusing the samples into a narrow region near the centerline of the microchannel, there is a wide range of applications [3–5] for increasing the detection efficiency in flow cytometry and throughput in particle sorting, as well as for protecting samples from unwanted interactions with the channel walls, which may cause shear or surface-induced damages to the sample. An efficient flow-focusing system should be able to push particles away from the walls of the channel and align them to move along defined flow paths. Focusing particles to a narrow stream, however, is not a trivial task. Due to the laminar nature of microfluidic flows, particles suspended in a fluid medium tend to follow the fluidic streamlines unless a lateral force moves them from their original paths [6]. Hydrodynamic forces have been commonly used to manipulate streamlines to guide particles to a confined region [7], which belongs to a passive focusing technique. However, many active focusing techniques, i.e., the use of external fields, have been applied to focused particles in microfluidic devices, such as acoustic, optical, and electrical focusing techniques, which typically require high power instruments (such as lasers or high voltage power supplies) [8].

Ferrofluids have demonstrated great potential for a variety of focusing of non-magnetic micro-particles/cells in microfluidics [9]. Compared to other technologies, magnetic focusing technology, which refers to the induced motion of particles in a non-uniform magnetic field, has distinct advantages, such as low cost, less sample consumption, no heating problem, and does not require expensive external systems as an aid [10]. Magnetic focusing technology, magnetophoresis, can be either positive or negative depending on whether the particle is more or less magnetizable than the suspending medium. Magnetic particles suspended in nonmagnetic solutions experience positive magnetophoresis and are directed along the magnetic field gradient toward a magnet [11]. In contrast, the non-magnetic particles suspended in the magnetic solutions are subjected to negative magnetophoresis and pushed away from the magnet [12]. The former phenomenon has been widely used to separate and sort cells or biomolecules in microfluidic devices by selectively labeling target cells with functionalized magnetic beads such as CellSearch technology [13], which is the only one approved by the US Food and Drug Administration (FDA).

In recent years, the use of negative magnetophoresis to focus particles in microchannels has attracted the attention of many scholars. Zhou et al. [14] developed a three-dimensional numerical model to simulate the transmission of diamagnetic particles during inertial focusing and magnetic separation in the entire microchannel and found that the predicted particle trajectories are roughly consistent with experimental observations. They further proposed a method for symmetrically embedding two repulsive permanent magnets around a linear rectangular microchannel in a microfluidic device based on polydimethylsiloxane (PDMS) to study the three-dimensional magnetic focusing of polystyrene particles in a ferrofluid [15]. Zhu et al. [16] have embedded permanent magnets in a PDMS-based microfluidic chip, in which magnets can be placed very close to a planar microchannel to enhance the magnetic field and field gradients. The non-magnetic particles can be focused continuously in a paramagnetic solution.

Since the end of an inclined magnet will form a high gradient region near the channel wall, an enhanced negative magnetophoresis will push the nonmagnetic particles toward the centerline by placing two opposite magnets on both sides of the channel. This situation is ideal for the magnetically induced flow focusing operation. There are few studies [17–19] on the ferrohydrodynamics with inclined magnetic fields, which focus on the effect of the flow patterns, inclination angle, and saturation magnetization on the heat transfer performance. To the best of our knowledge, no research explored the flow focusing of nonmagnetic particles by using the inclined magnetic fields. When the flow rate is also small, or the magnetic field is strong, the particles will be pushed back by the magnetic force before entering the magnetic field region, forming a circulation loop. It is necessary to consider the critical flow rate and the relative magnetic field parameters to control the particles accurately.

In this paper, we study the magnetically induced focusing of non-magnetic particles in ferrofluids under inclined magnetic fields. The effective focusing of the particles is achieved under different conditions including the rotation angle and critical Reynolds number. The paper is organized as follows: We first analyzed the various forces acting on the particles and determined the equation of motion of the particles. Then, we calculated the magnetic force that the particles are subjected to when the permanent magnets are placed in parallel or tilted. Again, we calculated the critical conditions, the critical magnetic field and flow rate, and compared them with the experimental results. Finally, we dimensioned the magnetic field and flow field and analyzed the focusing effectiveness of the particles at the exit of the channel.

2. Theoretical Analysis

Microfluidic magnetic focusing is a research field that involves the interaction between magnetism and fluid flow on a microscale [20]. In this section, we analyzed the vital forces that may affect the particle trajectory while passing through a microfluidic device. Non-magnetic microparticles in ferrofluids, under continuous flow conditions, are subject to a variety of forces, including magnetic force, viscous drag force, gravity/buoyancy, Brownian motion, the interaction between particles and surrounding medium, and interparticle effects, shown in Figure 1a,b. We first estimate the order of magnitude of each force to identify the dominant force for useful focusing.

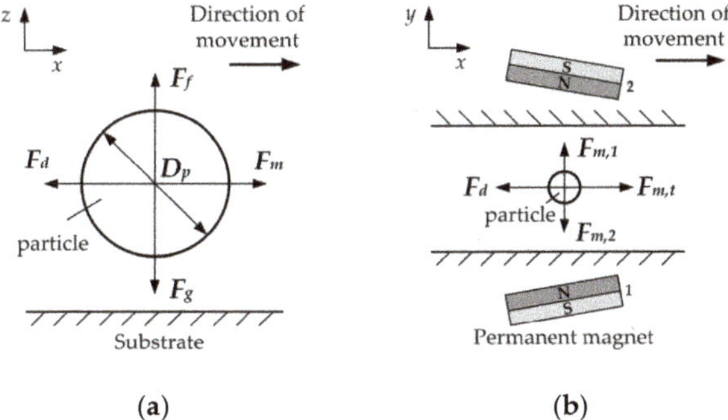

Figure 1. Schematic of the forces acting on the particle. (**a**) Forces analysis of a non-magnetic particle in the *x*, *z*-direction; (**b**) forces exerted on the particle when two permanent magnets are opposite in polarity.

2.1. Magnetic Force

Non-magnetic microparticle experience a negative magnetophoretic force, F_m, in a ferrofluid when subjected to a non-uniform magnetic field. The force can be written as [21]

$$F_m = -V_p \mu_0 (M \cdot \nabla) H, \tag{1}$$

where V_p is the volume of the individual microparticle; μ_0 is the magnetic permeability of the vacuum, and equal by $4\pi \times 10^{-7}$ H/m; the effective magnetization of the ferrofluids around the microparticles M (could be determined by the classical Langevin theory) is collinear with a static magnetic field H produced by the permanent magnet.

2.2. Viscous Drag Force

In low Reynolds number microfluidic systems, the hydrodynamic drag force, F_d, is related to the size and velocity of the particle, and the drag coefficient. We use the classical Stokes' formula to present the drag force:

$$F_d = 3\pi \eta D_p (u_p - u_f) C_D, \tag{2}$$

where η is the viscosity of suspensions; D_p is the diameter of the non-magnetic microparticles; u_p and u_f are the velocities of microparticles and suspensions, respectively. The coefficient, C_D, accounts for the increased fluid resistance when the particle moves near the microfluidic channel surface [22]:

$$C_D = \left[1 - \frac{9}{16}\left(\frac{D_p}{D_p + 2\Delta}\right) + \frac{1}{8}\left(\frac{D_p}{D_p + 2\Delta}\right)^3 - \frac{45}{256}\left(\frac{D_p}{D_p + 2\Delta}\right)^4 - \frac{1}{16}\left(\frac{D_p}{D_p + 2\Delta}\right)^5 \right]^{-1}, \tag{3}$$

where Δ defines the shortest distance between the particle surface and channel wall.

2.3. Gravity and Buoyancy

The gravity and buoyancy of the non-magnetic particles in ferrofluids can be written as a resultant force, F_n, as shown in the following equation:

$$F_n = \frac{\pi D_p^3}{6}\left(\rho_p - \rho_f\right)g, \tag{4}$$

where ρ_p and ρ_f are the density of the nonmagnetic particles and the magnetic fluid, respectively; g is the gravitational acceleration. For a 5 µm particle in EMG 607 ferrofluids (ρ_p = 1064 kg/m^3,

$\rho_f = 1100 \ \text{kg/m}^3$, $g = 9.8 \ \text{m/s}^2$), the calculated gravity and buoyancy are 6.98×10^{-2} pN and 7.20×10^{-2} pN, respectively, both of which are approximately two orders of magnitude smaller than magnetic and drag forces (~pN).

2.4. Brown Force

The phenomenon that suspended particles never stop moving irregularly is called Brownian motion. The Brownian force, F_B, can be expressed by the following formula: [23]

$$F_B = \zeta \sqrt{\frac{6\pi k_B \eta T D_p}{\Delta t}},\tag{5}$$

where k_B is the Boltzmann constant; T is the absolute temperature; Δt is the magnitude of the characteristic time step; the parameter, ζ, is the Gaussian random number with zero mean and unit variance. Brownian motion only affects the motion of the particle when the particle diameter is small enough (less than the critical diameter of the particle), or the F_m applied to the particle is weak [12]. In order to be able to determine the particle diameter that affects the Brownian motion, Gerber [24] et al. used the following formula to calculate the critical diameter of the particle:

$$|F|D_p \leq k_B T,\tag{6}$$

where $|F|$ is the resultant force of the non-magnetic particles. When the applied magnetic field is 1 T, the critical diameter of the Fe_4O_3 particles in water can be calculated as 40 nm according to the above formula.

Moreover, the interactions of the particle itself and with fluid usually could be neglected in a low concentration condition. Finally, only the magnetic force and drag force need to be considered in our study.

2.5. The Motion Equation of Particles

Figure 2 illustrates the focusing mechanism of non-magnetic particles in the magnet-microchannel system, where particles are pushed away from the high field region and concentrated along the centerline of the channel under the actions of two magnets with same opposite polarity. The motion of particles can be analyzed by considering dominant magnetic and hydrodynamic forces. Thus,

$$m_p \frac{du_p}{dt} = F_m + F_d.\tag{7}$$

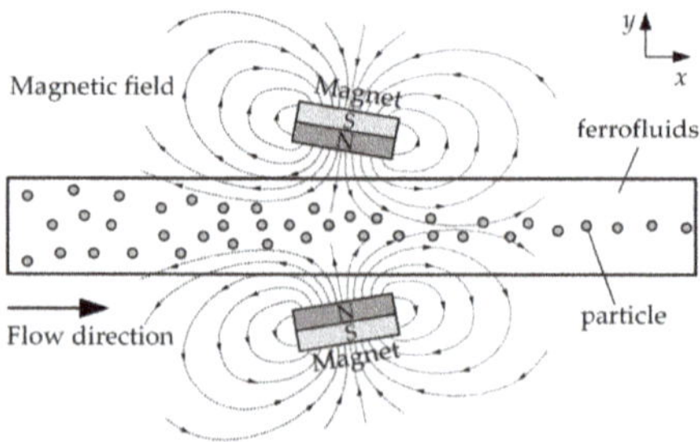

Figure 2. Schematic of the particles focusing mechanism. When two rectangular permanent magnets are placed at the center of a channel length with the same opposite polarity, and its direction of magnetization perpendicular to the channel wall, the magnet field magnetizes the ferrofluids within the microchannel and subsequently affects the particle motion.

3. Materials and Methods

Figure 3a shows a picture of the microfluidic chip adopted in our experiment. The straight channel was fabricated with PDMS (Sylgard 184, Dow Corning, Midland, MI, USA) by soft lithography [25]. Neodymium iron boron (NdFeB) permanent magnets were embedded on both sides of the channel with the same pole. The angle of rotation of the permanent magnet and the distance between the magnet and the channel can be changed during the experiment. The spatial position of the magnet and the channel is shown in Figure 3b and their specific sizes are given in the following section.

We used a commercial water-based magnetite ferrofluid (EMG 607, Ferrotec Corp., Tokyo, Japan) in our experiments. The volume fraction of magnetite particles for this ferrofluid is 1.8%. Initial magnetic susceptibility is 0.36; saturation magnetization is 7962 A/m; viscosity is 1.3×10^{-3} N·s/m^2. The particle solution was made by suspending 10.4 μm polystyrene particles (Bissler Corp., Tianjin, China) in ferrofluid at a volume ratio of 1:1. The dilute ferrofluid was prepared by mixing the original ferrofluid with pure water to a concentration of 0.022%. Tween 20 (Fisher Scientific Corp., Carlsbad, CA, USA) was added to the particle suspension at 0.1% by volume to minimize their aggregations and adhesions to channel walls.

As shown in Figure 3c, microparticles suspended in the diluted ferrofluid were driven through the microchannel by an infusion syringe pump (Harvard PHD 2000, Harvard Apparatus, Holliston, MA, USA). Two equal length tubes are inserted into the chip inlet and outlet, respectively, and bonded with glue. Before the experiment, the inlet tube was connected to the syringe, and the outlet tube was placed in the beaker. Particle motion was visualized using an inverted microscope (Nikon TE2000U, Tokyo, Japan) under bright-field illumination. Digital videos (at a time rate of around 12 frames per seconds) and images were recorded through a CCD camera (PixeLINK-B742F, Ottawa, ON, Canada) and post-processed using image analysis software (ImageJ; http://rsb.info.nih.gov/ij/).

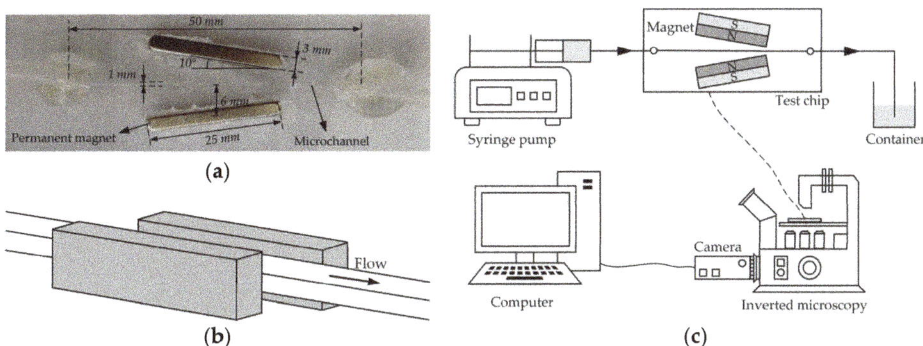

Figure 3. Experimental chip and system: (**a**) picture of the microfluidic chip with permanent magnets embedded in polydimethylsiloxane (PDMS); (**b**) schematic diagram of the spatial position of the channel and magnet, and note that the centerline of the permanent magnet is on the same plane as the centerline of the channel; (**c**) the experimental setup.

4. Results and Discussion

4.1. Magnetic Field

The system considered in the model consists of a microfluidic channel and a permanent magnet, in which dimensions of the channel and the magnet are labeled, as illustrated in Figure 4. The magnetic force experienced by a particle in the microchannel under a magnetic field can be expanded according to Equation (1),

$$F_{mx} = -V_p\mu_0 \left[M_x \frac{\partial H_x(x,y)}{\partial x} + M_y \frac{\partial H_x(x,y)}{\partial y} \right], \tag{8a}$$

$$F_{my} = -V_p \mu_0 \left[M_x \frac{\partial H_y(x,y)}{\partial x} + M_y \frac{\partial H_y(x,y)}{\partial y} \right]. \tag{8b}$$

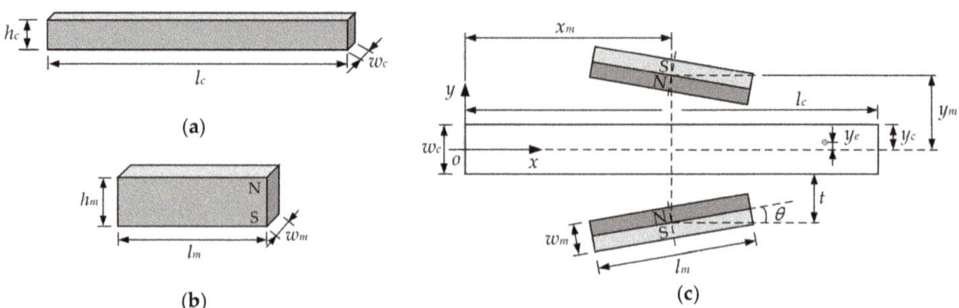

Figure 4. The dimensions of (**a**) the microfluidic channel and (**b**) the permanent magnets; (**c**) the relative locations of the microchannel and the magnets: l_c = 50 mm, w_c = 1 mm, h_c = 1 mm, l_m = 25 mm, w_m = 3 mm, h_m = 10 mm, t = 5.5 mm, x_m = 25 mm, y_m = 6 mm; x-y coordinate system is within the microchannel, with its origin at the center of the cross-section of the microchanel.

We use the Langevin equation to obtain the relationship between the saturation moment of the Fe_3O_4 nanoparticles M_d and the effective magnetization of the ferrofluids $\mathbf{M}$, as follows [21]:

$$\frac{\mathbf{M}}{\phi M_d} = coth(\alpha) - \frac{1}{\alpha}, \tag{9a}$$

$$\alpha = \frac{\pi \mu_0 M_d H d^3}{6 k_B T}, \tag{9b}$$

where d = 10 nm is the average diameter of the Fe_3O_4 nanoparticles; ϕ is the particle volume concentration; $M_d = 4.423 \times 10^5$ A/m can be calculated from the manufacturer-provided saturation magnetization of ferrofluids. Moreover, the analytical expressions for x- and y-component of the effective magnetization can be expressed as

$$M_x = M \frac{H_x}{H}, \tag{10a}$$

$$M_y = M \frac{H_y}{H}. \tag{10b}$$

The analytical expressions for the x- and y-component of the magnetic field strength for single rectangular magnet are given by Equation (11a,b) [26].

$$H_x(x,y) = \frac{M_s}{4\pi} \left\{ ln\left[\frac{(x+0.5l_m)^2 + (y-0.5w_m)^2}{(x+0.5l_m)^2 + (y+0.5w_m)^2} \right] - ln\left[\frac{(x-0.5l_m)^2 + (y-0.5w_m)^2}{(x-0.5l_m)^2 + (y+0.5w_m)^2} \right] \right\}, \tag{11a}$$

$$H_y(x,y) = \frac{M_s}{2\pi} \left\{ tan^{-1}\left[\frac{w_m(x+0.5l_m)}{(x+0.5l_m)^2 + y^2 - 0.25w_m^2} \right] - tan^{-1}\left[\frac{w_m(x-0.5l_m)}{(x-0.5l_m)^2 + y^2 - 0.25w_m^2} \right] \right\}. \tag{11b}$$

The expressions for the magnetic field gradients are

$$\frac{\partial H_x(x,y)}{\partial x} = \frac{M_s}{2\pi} \left[\frac{x+0.5l_m}{(x+0.5l_m)^2 + (y-0.5w_m)^2} - \frac{x+0.5l_m}{(x+0.5l_m)^2 + (y+0.5w_m)^2} - \frac{x-0.5l_m}{(x-0.5l_m)^2 + (y-0.5w_m)^2} + \frac{x-0.5l_m}{(x-0.5l_m)^2 + (y+0.5w_m)^2} \right], \tag{12a}$$

$$\frac{\partial H_x(x,y)}{\partial y} = \frac{M_s}{2\pi}\left[\frac{y-0.5w_m}{(x+0.5l_m)^2+(y-0.5w_m)^2}-\frac{y-0.5w_m}{(x-0.5l_m)^2+(y-0.5w_m)^2}-\frac{y+0.5w_m}{(x+0.5l_m)^2+(y+0.5w_m)^2}+\frac{y+0.5w_m}{(x-0.5l_m)^2+(y+0.5w_m)^2}\right],\tag{12b}$$

$$\frac{\partial H_y(x,y)}{\partial x} = \frac{M_s}{2\pi}\left[\frac{w_m\left[y^2-(x+0.5l_m)^2-0.25w_m^2\right]}{\left[(x+0.5l_m)^2+y^2-0.25w_m^2\right]^2+w_m^2(x+0.5l_m)^2}-\frac{w_m\left[y^2-(x-0.5l_m)^2-0.25w_m^2\right]}{\left[(x-0.5l_m)^2+y^2-0.25w_m^2\right]^2+w_m^2(x-0.5l_m)^2}\right],\tag{12c}$$

$$\frac{\partial H_y(x,y)}{\partial y} = \frac{M_s}{\pi}\left[\frac{w_m y(x-0.5l_m)}{\left[(x-0.5l_m)^2+y^2-0.25w_m^2\right]^2+w_m^2(x-0.5l_m)^2}-\frac{w_m y(x+w)}{\left[(x+0.5l_m)^2+y^2-0.25w_m^2\right]^2+w_m^2(x+0.5l_m)^2}\right],\tag{12d}$$

where $M_s = B_r/\mu_0$ is magnetization of the permanent magnet, B_r is remanent magnetization of the permanent magnets.

The above formula is based on the origin of the permanent magnet as the coordinate origin, and the permanent magnet is placed vertically, that is, the magnetization direction is the y-direction, as shown in Figure 4b, the N-pole of the permanent magnet is on the upper side, and the S-pole is on the lower side. However, we choose the center at the entrance of the channel as the origin of the coordinates. Therefore, coordinate transformation (translation and rotation) is necessary. A rotation matrix is used to perform a rotation of points in the xy-Cartesian plane counter-clockwise through an angle, about the origin of the coordinate system. The coordinates after translation and rotation are obtained by using matrix multiplication. The coordinate translation is $(x\text{-}x_m, y+y_m)$, and the rotation angle is θ ($0° \leq \theta \leq 90°$) for the permanent magnet below the channel (Equation (13a)), while for the permanent magnet above the channel, the coordinate translation is $(x\text{-}x_m, y\text{-}y_m)$ and the rotation angle is $\pi\text{-}\theta$, Equation (13b). Note that the subscript 1 (and subscript 2) in the following matrix equations represent the correlation calculation of the permanent magnets below (and above) the channel.

$$\begin{bmatrix} x_1' \\ y_1' \end{bmatrix} = \begin{bmatrix} \cos\theta & \sin\theta \\ -\sin\theta & \cos\theta \end{bmatrix}\begin{bmatrix} x-x_m \\ y+y_m \end{bmatrix},\tag{13a}$$

$$\begin{bmatrix} x_2' \\ y_2' \end{bmatrix} = \begin{bmatrix} \cos(\pi-\theta) & \sin(\pi-\theta) \\ -\sin(\pi-\theta) & \cos(\pi-\theta) \end{bmatrix}\begin{bmatrix} x-x_m \\ y-y_m \end{bmatrix}.\tag{13b}$$

Substituting the Equation (13a,b) into the Equation (11a,b), respectively, the magnetic field strength after the rotation of the permanent magnet can be obtained, and then the magnetic field strength in the coordinate system of Figure 4c can be obtained by conversion, see Equation (14a,b).

$$\begin{bmatrix} H_{x_1,r}(x,y) \\ H_{y_1,r}(x,y) \end{bmatrix} = \begin{bmatrix} \cos\theta & -\sin\theta \\ \sin\theta & \cos\theta \end{bmatrix}\begin{bmatrix} H_{x_1'}(x_1',y_1') \\ H_{y_1'}(x_1',y_1') \end{bmatrix},\tag{14a}$$

$$\begin{bmatrix} H_{x_2,r}(x,y) \\ H_{y_2,r}(x,y) \end{bmatrix} = \begin{bmatrix} \cos(\pi-\theta) & -\sin(\pi-\theta) \\ \sin(\pi-\theta) & \cos(\pi-\theta) \end{bmatrix}\begin{bmatrix} H_{x_2'}(x_2',y_2') \\ H_{y_2'}(x_2',y_2') \end{bmatrix}.\tag{14b}$$

In the same way, we can get the rotated magnetic field gradients in the coordinate system of Figure 4c, see Equation (15a,b).

$$\begin{bmatrix} \frac{\partial H_{x_1,r}(x,y)}{\partial x} & \frac{\partial H_{x_1,r}(x,y)}{\partial y} \\ \frac{\partial H_{y_1,r}(x,y)}{\partial x} & \frac{\partial H_{y_1,r}(x,y)}{\partial y} \end{bmatrix} = \begin{bmatrix} \cos\theta & -\sin\theta \\ \sin\theta & \cos\theta \end{bmatrix}\begin{bmatrix} \frac{\partial H_{x_1'}(x_1',y_1')}{\partial x_1'} & \frac{\partial H_{x_1'}(x_1',y_1')}{\partial y_1'} \\ \frac{\partial H_{y_1'}(x_1',y_1')}{\partial x_1'} & \frac{\partial H_{y_1'}(x_1',y_1')}{\partial y_1'} \end{bmatrix}\begin{bmatrix} \cos\theta & \sin\theta \\ -\sin\theta & \cos\theta \end{bmatrix},\tag{15a}$$

$$\begin{bmatrix} \dfrac{\partial H_{x_2,r}(x,y)}{\partial x} & \dfrac{\partial H_{x_2,r}(x,y)}{\partial y} \\ \dfrac{\partial H_{y_2,r}(x,y)}{\partial x} & \dfrac{\partial H_{y_2,r}(x,y)}{\partial y} \end{bmatrix} = \begin{bmatrix} \cos(\pi-\theta) & -\sin(\pi-\theta) \\ \sin(\pi-\theta) & \cos(\pi-\theta) \end{bmatrix} \begin{bmatrix} \dfrac{\partial H_{x_2'}(x_2',y_2')}{\partial x_2'} & \dfrac{\partial H_{x_2'}(x_2',y_2')}{\partial y_2'} \\ \dfrac{\partial H_{y_2'}(x_2',y_2')}{\partial x_2'} & \dfrac{\partial H_{y_2'}(x_2',y_2')}{\partial y_2'} \end{bmatrix} \begin{bmatrix} \cos(\pi-\theta) & \sin(\pi-\theta) \\ -\sin(\pi-\theta) & \cos(\pi-\theta) \end{bmatrix}. \tag{15b}$$

Therefore, the magnetic force acting in the x- and y-directions on the particle can be calculated by Equation (8a,b). Then we add the magnetic forces generated by the two permanent magnets at the same position to obtain the final magnetic force in the x- and y-directions.

$$F_{mx} = F_{mx_1} + F_{mx_2}, \tag{16a}$$

$$F_{my} = F_{my_1} + F_{my_2}. \tag{16b}$$

4.1.1. The Vertical Magnet Filed

The vertical magnetic field means that two permanent magnets are placed in parallel with the same polarity, i.e., $\theta = 0$. Based on the above analysis, we can calculate the magnetic field strength and the magnetic force acting on the particle at the center of the channel, i.e., $y = 0$.

Figure 5a,b shows the magnetic field strength and the magnetic force generated by two permanent magnets from the channel inlet ($x = 0$ mm) to the outlet ($x = 50$ mm) at the center of the channel ($y = 0$ mm), respectively. Although the polarities of the first and second permanent magnets are opposite, the resulting magnetic field strength in the x-direction is the same, with a maximum of ± 73.52 kA/m on the left (right) sides and a minimum of 0 kA/m in the middle. The value of the magnetic field strength in the y-direction is the opposite number, with a minimum of 0 kA/m on both sides and a saddle-shaped change in the middle. Therefore, the total magnetic field strength is twice the original value for the x-direction and 0 kA/m for the y-direction, as shown in Figure 5a.

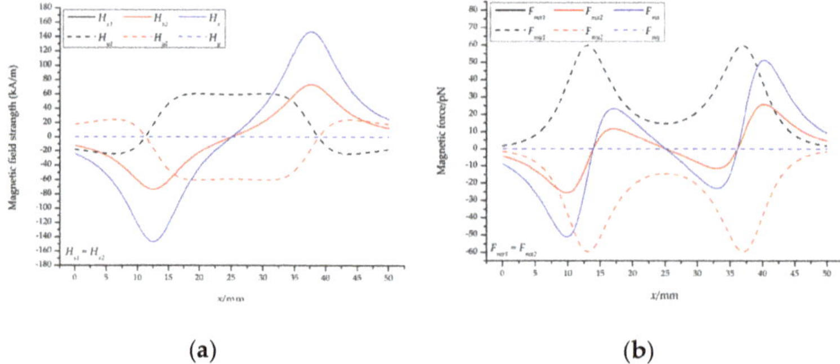

(a)   (b)

Figure 5. The results of vertical magnetic field: (**a**) The magnetic field strength and (**b**) the magnetic force acting on the particles at the center of the channel. Note that the residual magnetization of a single permanent magnet is 1.2 T.

The trend of the corresponding magnetic force is also the same, that is, the values in the x-direction (y-direction) are the same (opposite), as shown in Figure 5a. The magnetic force in the y-direction at the center of the channel is 0 N, indicating that the particles at the center are only related to the magnetic force in the x-direction, and the particles on both sides of the center line will move toward the center. It can be seen from the total magnetic force curve in the x-direction that the particles will decelerate (maximum: -25.57 pN) and accelerate (maximum: 11.53 pN) when moving at the center of the channel, and then decelerate (maximum: -11.53 pN) and accelerate (maximum: 25.57 pN).

These facts indicate that the strength of the magnetic field of the permanent magnet is proportional to the magnetic force acting on the particles; the particles at both ends of the permanent magnet are subjected to the largest magnetic force and the smallest in the middle; the particles are subjected to a negative magnetic force when initially entering the magnetic field, which will prevent the particles from continuing to move forward, after which the particles are substantially subjected to a positive magnetic force, causing the particles to accelerate.

4.1.2. The Inclined Magnet Field

The inclined magnetic field means that the two permanent magnets with the same polarity are placed at an angle θ ($0° \leq \theta \leq 90°$) on both sides of the channel. Analogous to the vertical magnetic field, we calculated the magnetic field strength and magnetic force at the center of the channel when the angle changes from $10°$ to $50°$, as shown in Figure 6. Note that Figure 6a–d shows the magnetic field strength and the magnetic force of the permanent magnet below the channel; the permanent magnet above the channel is only a sign change, which can be referred to as a vertical magnetic field, which is not shown here.

A slight change in angle will cause a large change in magnetic field strength and magnetic force. Regardless of the x-direction or the y-direction, the magnetic field strength at the center of the channel decreases on the left side of the permanent magnet and increases on the right side, as shown in Figure 6a,b. For example, the magnetic field strength in the x-direction of the left side is reduced from $|-73.52|$ kA/m to $|-53.63|$ kA/m, and the right side is increased from 73.52 kA/m to 100.74 kA/m when the angle $\theta = 10°$. However, we found that there is a transition on the right side when the angle is increased to $30°$; increasing the angle to $40°$, the magnetic field strength in the x-direction turns to a negative value, and the y-direction decreases; and then increases the angle to $50°$, the magnetic field strength in x- and y-direction continues to decrease.

The trend of magnetic force is roughly the same as the strength of the magnetic field: When $\theta = 30°$, there is also a transition, as shown in Figure 6c,d. Figure 6e,f show the total magnetic field strength and the total magnetic force at the center of the channel, respectively. When the permanent magnet is rotated by a certain angle, the magnetic force on the left side decreases and the right side increases. For example, when the angle $\theta = 10°$, the left magnetic force decreases from $|-25.57|$ pN to $|-13.48|$ pN, and the right side increases from 25.57 pN to 68.65 pN.

Therefore, when the permanent magnet rotates at a certain angle, the magnetic field strength and magnetic force will also change. This does not mean that the angle can be arbitrarily rotated, because the magnetic field strength and magnetic force have a transition when the angle is rotated to $30°$, which is not conducive to particle focusing. It is appropriate to choose a rotation angle of $0° < \theta < 30°$.

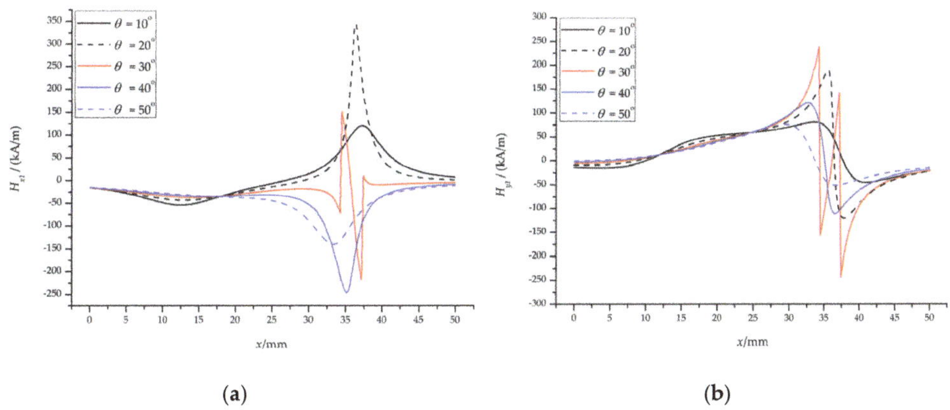

(a)

(b)

Figure 6. *Cont.*

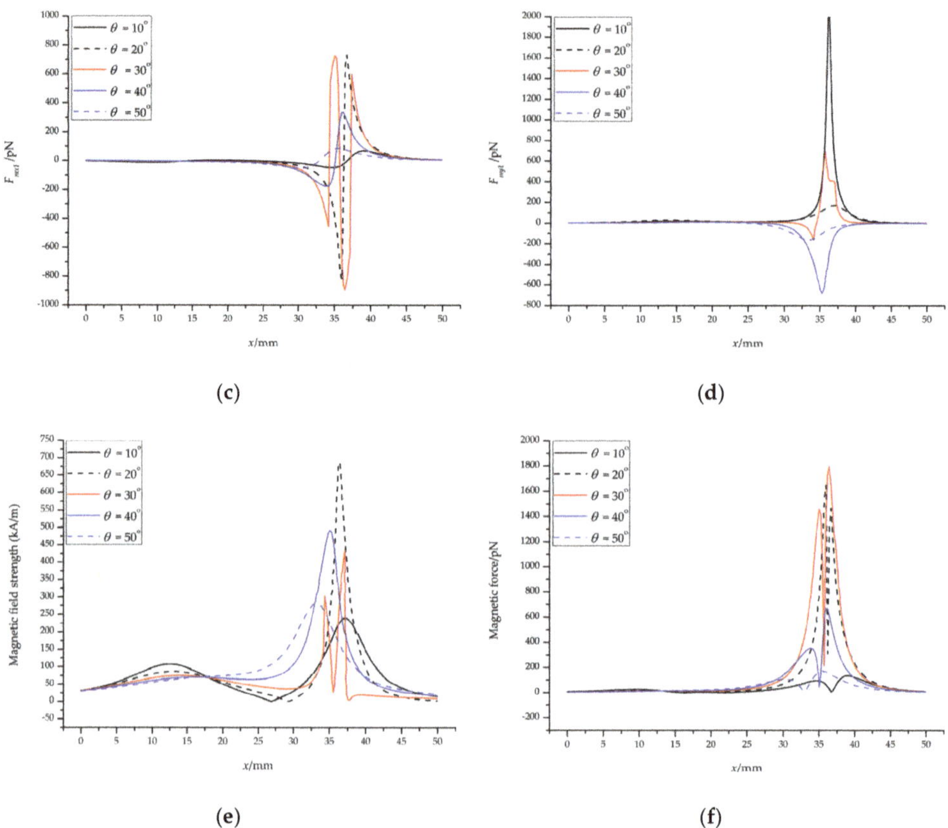

Figure 6. The results of the inclined magnetic field: The magnetic field strength (**a**) in the *x*-direction and (**b**) in the *y*-direction; the magnetic force (**c**) in the *x*-direction and (**d**) in the *y*-direction; (**e**) the total magnetic field strength and (**f**) the total magnetic force at the center of the channel.

We also calculated the magnetic force in the *y*-direction of the half channel from $y = 100$ μm to $y = 400$ μm ($y = 0$ at the centerline of the channel, see Figure 4c), as shown in Figure 7. The maximum magnetic forces in the *y*-direction are -4.62 pN ($y = 100$ μm), -9.27 pN ($y = 200$ μm), -13.95 pN ($y = 300$ μm), and -18.68 pN ($y = 400$ μm), respectively, for the vertical magnetic field ($\theta = 0°$, $Br = 1.2$ T), as shown in Figure 7a. For the inclined magnetic field ($\theta = 10°$, $Br = 1.2$ T), the maximum magnetic forces are -22.03 pN, -44.36 pN, -67.27 pN, -91.11 pN, respectively, when *y* is 100 μm, 200 μm, 300 μm, and 400 μm, respectively, as shown in Figure 7b.

These values are negative because the particles are subjected to negative magnetophoresis (permanent magnet repulsive force) and move toward the centerline of the channel. For the lower half of the channel, the curve is opposite to Figure 7, with absolute values equal. It is because of the negative magnetophoresis effect of the two permanent magnets that the particles are focused in the center of the channel. And when the permanent magnet is inclined at a certain angle, the negative magnetophoresis effect is enhanced, which is more conducive to the particles focusing.

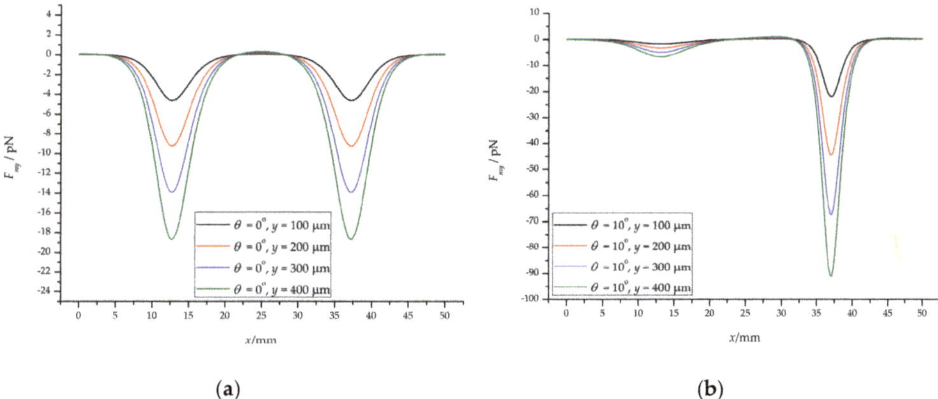

Figure 7. The results of magnetic force in the *y*-direction (half channel): (**a**) the vertical magnetic field; (**b**) the inclined magnetic field.

4.2. Flow Field

The fluid flow pattern needs to be determined to calculate the fluid flow rate in the channel. The flow state of the fluid can be judged according to the Reynolds number calculation formula:

$$Re = \frac{\rho U D_h}{\eta} = \frac{2\rho Q}{\eta(w_c + h_c)},\tag{17}$$

where U is the average fluid velocity, $D_h = 2w_c h_c / (w_c + h_c)$ is the hydraulic diameter of the rectangular microchannel, and Q is the volumetric flow rate. Normally, $\rho = 10^3$ kg/m^3, $U = 10^{-3}$ m/s, $D_h = 10^{-3}$ m, $\eta = 10^{-3}$ N·s/m^2, then $Re = 1$, which belongs to laminar flow. In the case of laminar flow, the formula for calculating the fluid velocity in the channel can be obtained [27,28].

$$u_f = \left(\frac{\Delta p}{l_c}\right)\left(\frac{4h_c^3}{\eta\pi^3}\right)\sum_{n=0}^{\infty}\frac{(-1)^n}{(2n+1)^3}\times\left[1-\frac{cosh\frac{(2n+1)\pi y}{h_c}}{cosh\frac{(2n+1)\pi w_c}{2h_c}}\right]\times cosh\frac{(2n+1)\pi z}{h_c},\tag{18}$$

where Δp is the pressure drop of the fluid along the length of the channel; n is an integer. Here, we introduce the channel aspect ratio $\varepsilon = h_c/w_c$, and studies have shown that when $1 < \varepsilon < 2$, the calculation error will be controlled within 1%, and if the ratio of channel aspect ratio is larger, the calculation error will also increase. For the small Reynolds number flow, in order to control the calculation error in Equation (18), only when $n = 1$ and 2, the calculation result can represent the flow velocity in the channel, and the final result is more accurate, so Equation (18) can be rewritten as follows:

$$u_f = \left(\frac{\Delta p}{l_c}\right)\left(\frac{4h_c^3}{\eta\pi^3}\right)\left\{\left[1-\frac{cosh\frac{\pi y}{h_c}}{cosh\frac{\pi w_c}{2h_c}}\right]\times cosh\frac{\pi z}{h_c}-\frac{1}{27}\left[1-\frac{cosh\frac{3\pi y}{h_c}}{cosh\frac{3\pi w_c}{2h_c}}\right]\times cosh\frac{3\pi z}{h_c}\right\},\tag{19}$$

where the pressure drop Δp is a function of the volumetric flow rate Q of the fluid in the channel inlet and can be expressed by the following formula [29]:

$$\Delta p = \frac{Ql_c\eta\pi^4}{8w_c h_c^3}\times\left\{\left[1-\frac{2h_c}{\pi w_c}tanh\left(\frac{\pi w_c}{2h_c}\right)\right]+\frac{1}{81}\left[1-\frac{2h_c}{3\pi w_c}tanh\left(\frac{3\pi w_c}{2h_c}\right)\right]\right\}^{-1}.\tag{20}$$

Substituting Equation (20) into Equation (19) and arranging it, the final formula for the flow velocity in the channel is:

$$u_f = \frac{Q\pi}{2w_c h_c} \times \left\{ \left[1 - \frac{2h_c}{\pi w_c} tanh\left(\frac{\pi w_c}{2h_c}\right) \right] + \frac{1}{81}\left[1 - \frac{2h_c}{3\pi w_c} tanh\left(\frac{3\pi w_c}{2h_c}\right) \right] \right\}^{-1} \times$$
$$\left\{ \left[1 - \frac{cosh\frac{\pi y}{h_c}}{cosh\frac{\pi w_c}{2h_c}} \right] \times cos\frac{\pi z}{h_c} - \frac{1}{27}\left[1 - \frac{cosh\frac{3\pi y}{h_c}}{cosh\frac{3\pi w_c}{2h_c}} \right] \times cos\frac{\pi z}{h_c} \right\}. \tag{21}$$

In the laminar flow state, the motion equation of the particles, Equation (7), can be simplified to

$$F_m + F_d = 0. \tag{22}$$

Therefore, the flow rate of the particles can be expressed as

$$u_{px} = u_{mx} + u_{fx} = \frac{F_{mx}}{3\pi\eta D_p C_D} + u_f, \tag{23a}$$

$$u_{py} = u_{my} + u_{fy} = \frac{F_{my}}{3\pi\eta D_p C_D}. \tag{23b}$$

Figure 8a,b shows the flow velocity distribution of the ferrofluids and particles, and the corresponding drag coefficient. The velocity distribution of the *xy* cross-section in the rectangular channel is parabolic, which is approximately the same as the velocity distribution of the circular channel (Figure 8a). The velocity distribution of the particles is closely related to the magnitude of the magnetic force, as shown in Figure 8b. The minimum flow rate is 4.87×10^{-4} m/s on the left side of the permanent magnet, and the maximum flow rate is 9.02×10^{-4} m/s on the right side, which is consistent with the above analysis. For the *y*-direction, the effect is to push the particles toward the center of the channel. The specific analysis is given in Section 4.4.

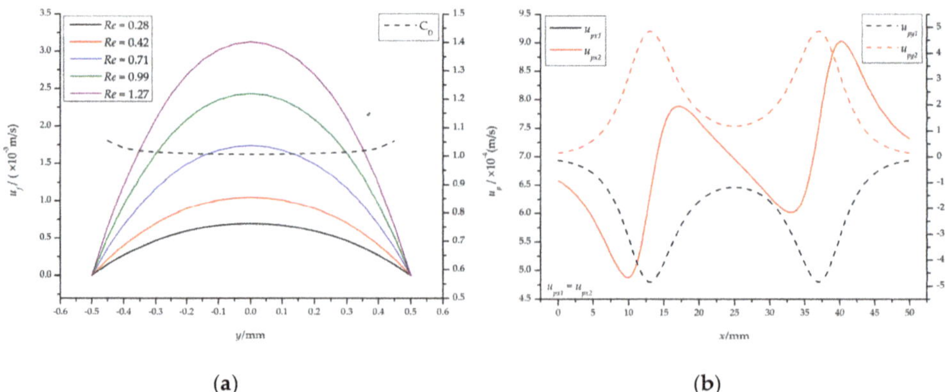

Figure 8. The results of flow field: (**a**) The flow rate of the ferrofluids in the channel at different inlet flow rates and drag coefficient along the *y*-direction; (**b**) the velocity distribution of particles in the center of the channel at $Q = 20$ μL/min ($Re = 0.28$).

4.3. Magnetic Field and Flow Field Threshold

From Figure 5b we know that F_{mx} has a negative peak on the left side of the permanent magnet, which has a large effect on u_{mx} from Equation (23a). Once $u_{mx} < u_{fx}$, i.e., $u_{px} < 0$, the particles will move in the negative *x*-direction, forming a loop. Therefore, in order to ensure that the particles can effectively enter the magnetic field and focus to the center of the channel, it is necessary to calculate the critical value of u_{fx} (Q). From the analysis in Section 4.1, we know that $0° < \theta < 30°$, so we calculated the critical values at $0°$, $10°$, and $20°$.

Figure 9 shows the critical values in different magnetic fields (Br) and different angles (θ). Under the same magnetic field, the critical value increases with the increase of the angle, especially when $\theta = 20°$, the amplitude is larger, as shown in Figure 9a,b. For $\theta = 20°$, the critical value is 499.78 μL/min ($Re_c = 7.05$) at $Br = 1.2$ T (Figure 9a) and has reached 1021.83 μL/min ($Re_c = 14.41$) at $Br = 2.4$ T (Figure 9b) in 36.42 mm of the channel. These values are quite large for microchannels, so we do not consider this case.

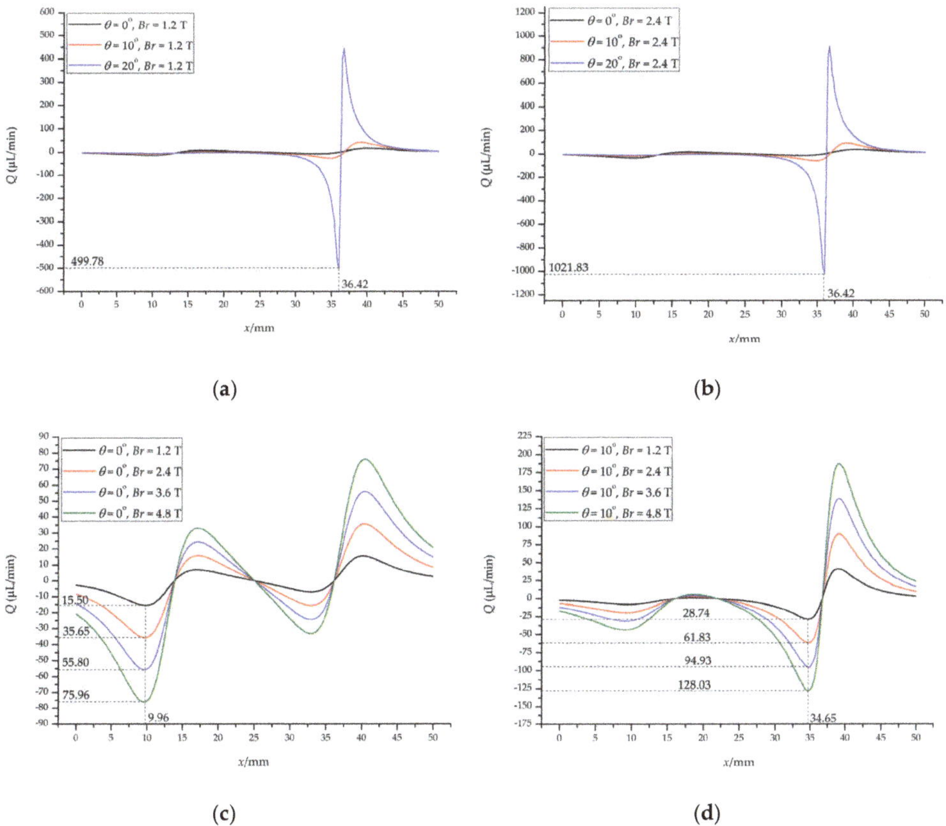

Figure 9. The threshold of magnetic field and flow field: the corresponding critical value of the permanent magnet after rotating at different angles for (**a**) $Br = 1.2$ T and (**b**) $Br = 2.4$ T; the corresponding critical value of (**c**) the vertical magnetic field and (**d**) the inclined magnetic field under different Br.

Figure 9c shows the critical value under the vertical magnetic field ($\theta = 0°$). The critical value is 15.50 μL/min ($Re_c = 0.22$) for $Br = 1.2$ T and 35.65 μL/min ($Re_c = 0.50$) for $Br = 2.4$ T; when $Br = 3.6$ T and 4.8 T, the critical value is 55.80 μL/min ($Re_c = 0.79$) and 75.96 μL/min ($Re_c = 1.07$), respectively. In both cases, it is located at 9.96 mm of the channel, which means that as long as the inlet flow rate Q exceeds this value, the particles will smoothly enter the magnetic field instead of forming a loop. Figure 9d gives the critical value of the vertical magnetic field ($\theta = 10°$). The critical value for this case is generated at 34.65 mm of the channel compared to the vertical magnetic field. For $Br = 1.2$ T and 2.4 T, the critical value is 28.74 μL/min ($Re_c = 0.41$) and 61.83 μL/min ($Re_c = 0.87$), respectively.

The critical value is 94.93 μL/min (Re_c = 1.34) and 128.03 μL/min (Re_c = 1.81) when Br = 3.6 T and 4.8 T, respectively. These values are greater than the vertical magnetic field at the same Br.

These results show that both the angle of the permanent magnet and the residual magnetization of the permanent magnet influence the critical flow rate, which in turn affects the particles focusing. Although the critical flow rate is exceeded, the flow rate at the end of the channel is quite large, which requires the consideration of the effect of fluid inertia on the particles. In short, we need to consider all aspects to achieve effective control of particles/cells to meet a variety of application needs.

Figure 10 shows the pictures we obtained through experiments. In the case of a vertical magnetic field, a loop is formed when Q is less than the critical flow rate (Q_c); and an increase in the strength of the magnetic field will cause more particles to bounce back, forming a thicker loop, as shown in Figure 10a,b. The same is true for the inclined magnetic field (Figure 10c,d), and the particles in the loop of the vertical magnetic field are less than the inclined magnetic field under the same Br.

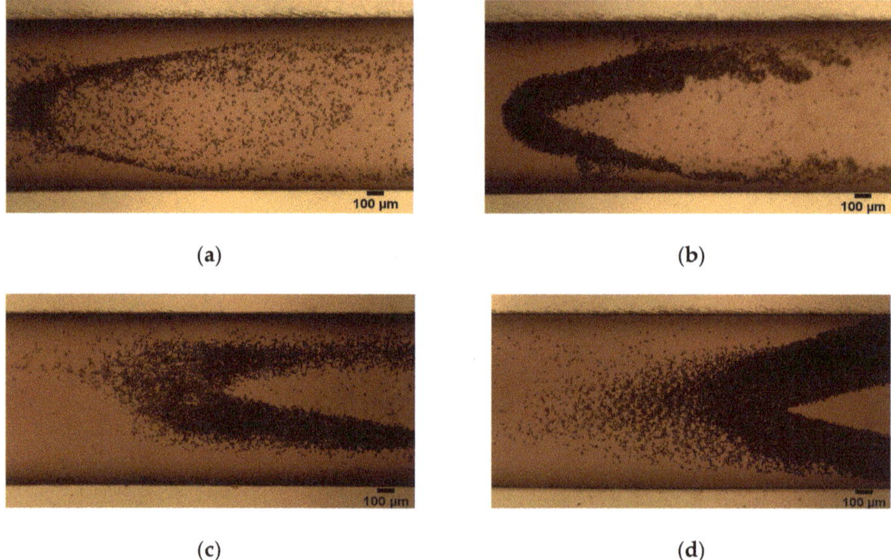

Figure 10. A loop formed below the critical flow rate. The vertical magnetic field (θ = 0°): (a) Br = 1.2 T, Q = 10 μL/min (Re = 0.01) and (b) Br = 2.4 T, Q = 30 μL/min (Re = 0.42); the inclined magnetic field (θ = 10°): (c) Br = 1.2 T, Q = 20 μL/min (Re = 0.28) and (d) Br = 2.4 T, Q = 50 μL/min (Re = 0.71).

The particles overcome the influence of the negative magnetic force and smoothly enter the magnetic field and focus when $Q > Q_c$. We took a picture of the focusing of particles near the exit of the channel, as shown in Figure 11. The experimental results in Figures 10 and 11 demonstrate that the critical flow rate determined by our theoretical analysis is correct.

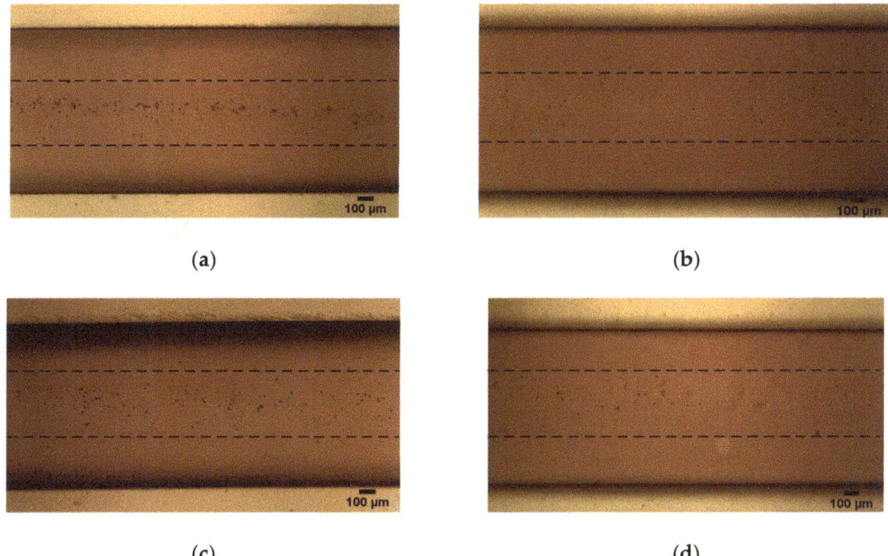

(a)

(b)

(c)

(d)

Figure 11. Focusing near the exit of the channel under the vertical magnetic field ($\theta = 0°$): (**a**) $Br = 1.2$ T, $Q = 20$ μL/min ($Re = 0.28$) and (**b**) $Br = 2.4$ T, $Q = 40$ μL/min ($Re = 0.56$); the inclined magnetic field ($\theta = 10°$): (**c**) $Br = 1.2$ T, $Q = 30$ μL/min ($Re = 0.42$) and (**d**) $Br = 2.4$ T, $Q = 70$ μL/min ($Re = 0.99$).

4.4. Focusing Effectiveness

The particles can be focused when $Q > Q_c$ ($Re > Re_c$). To facilitate a quantitative comparison of non-magnetic particle focusing in ferrofluids under different Q, we define a dimensionless number, focusing effectiveness, to evaluate the particle focusing performance,

$$\text{Focusing effectiveness} = \frac{\text{Half width of the focused particles stream}}{\text{Half width of the channel}} = \frac{y_e}{y_c}, \tag{24}$$

where the width of the microchannel was fixed at 500 μm in all tests of this work and the width of the focused particle stream was measured near the exit of the channel on the experimental images by ImageJ.

Figure 12a shows that an increase in flow rate results in a decrease in focusing effectiveness whether θ varies from $0°$ to $10°$ or Br varies from 1.2 T to 2.4 T. For $\theta = 0°$ and $Br = 1.2$ T, the minimum value of y_e is 263.03 μm ($Q = 20$ μL/min, $Re = 0.28$), the corresponding effectiveness is 52.61%; the maximum value is 473.94 μm ($Q = 80$ μL/min, $Re = 1.13$), and the corresponding effectiveness is 5.21%; for $\theta = 0°$ and $Br = 2.4$ T, the minimum and maximum value of y_e is 106.67 μm ($Q = 40$ μL/min, $Re = 0.56$) and 448.89 μm ($Q = 100$ μL/min, $Re = 1.41$), respectively, and the corresponding effectiveness is 78.67% and 10.22%, respectively, as shown in Figure 12b. For $\theta = 5°$ and $Br = 1.2$ T, the minimum value of y_e is 228.62 μm ($Q = 30$ μL/min, $Re = 0.42$), the corresponding effectiveness is 54.28%; the maximum value is 455.59 μm ($Q = 90$ μL/min, $Re = 1.27$), and the corresponding efficiency is 8.88%; for $\theta = 5°$ and $Br = 2.4$ T, the minimum and maximum value of y_e is 217.68 μm ($Q = 70$ μL/min, $Re = 0.99$) and 492.03 μm ($Q = 130$ μL/min, $Re = 1.83$), respectively, and the corresponding effectiveness is 56.46% and 1.60%, respectively. For $\theta = 10°$ and $Br = 1.2$ T, the minimum value of y_e is 183.58 μm ($Q = 30$ μL/min, $Re = 0.42$), the corresponding effectiveness is 63.29%; the maximum value is 454.11 μm ($Q = 90$ μL/min, $Re = 1.27$), and the corresponding efficiency is 9.18%; for $\theta = 10°$ and $Br = 2.4$ T, the minimum and maximum value of y_e is 179.72 μm ($Q = 70$ μL/min, $Re = 0.99$) and 473.18 μm ($Q = 130$ μL/min, $Re = 1.83$), respectively, and the corresponding effectiveness is 64.06% and 5.36%, respectively.

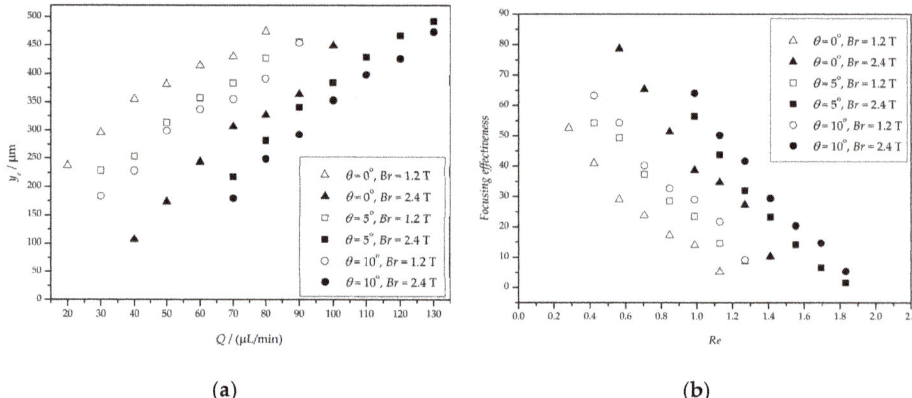

Figure 12. The results of focusing particles at different flow rates. (**a**) The width of the focused particle stream; (**b**) the focusing effectiveness.

These facts indicate that the focusing effectiveness is better when Q is near Q_c and the focusing effectiveness of $Br = 1.2$ T is less than $Br = 2.4$ T when θ is the same. Therefore, choosing the right angle and flow rate is critical to control the particles/cells accurately.

5. Conclusions

The focusing of biological and synthetic particles in microfluidic devices is a crucial step for the construction of many microstructured materials as well as for medical applications. We analyzed the performance of the magnetic focusing of the particles under a vertical magnetic field and an inclined magnetic field. We found that when the rotation angle of the permanent magnet exceeds 30°, an inclined magnetic field is unsuitable for particle focusing. We also calculated the effect of the flow field on the particle focusing, i.e., the critical flow rate. The calculations show that when the rotation angle exceeds 20°, the critical minimum flow rate exceeds 500 μL/min, which is quite large for the microchannel. Therefore, the optimum value of the rotation angle of the magnets should be between 0° and 20°, and 10° is preferred. The results also show that the critical flow rate is 15.5 μL/min and 35.65 μL/min for $\theta = 0°$, $Br = 1.2$ T and 2.4 T, respectively; for $\theta = 10°$, $Br = 1.2$ T and 2.4 T, the critical flow rate is 28.74 μL/min and 61.83 μL/min, respectively. These relationships have been verified with experimental results, which agree quantitatively with the predictions.

Author Contributions: Conceptualization, Y.H.; formal analysis, L.L.; investigation, L.L.; methodology, L.L.; supervision, Y.H.; writing—original draft, L.L.; writing—review and editing, Y.H.

Funding: This work was funded by the National Natural Science Foundation of China (grant number 11502102).

Conflicts of Interest: The authors declare no conflict of interest.

References

1. Tang, W.; Tang, D.; Ni, Z.; Xiang, N.; Yi, H. A microfluidic impedance cytometer with inertial focusing and liquid electrodes for high-throughput cell counting and discrimination. *Anal. Chem.* **2017**, *89*, 3154–3161. [CrossRef] [PubMed]
2. Xiao, W.; Matthew, Z.; Chia-Chi, H.; Necati, K.; Ian, P. Single stream inertial focusing in a straight microchannel. *Lab Chip* **2015**, *15*, 1812–1821. [CrossRef]
3. Huang, S.; He, Y.Q.; Jiao, F. Advances of particles/cells magnetic manipulation in microfluidic chips. *Chin. J. Anal. Chem.* **2017**, *45*, 1238–1246. [CrossRef]
4. Shields, C.W., IV; Reyes, C.D.; López, G.P. Microfluidic cell sorting: A review of the advances in the separation of cells from debulking to rare cell isolation. *Lab Chip* **2015**, *15*, 1230–1249. [CrossRef] [PubMed]

5. Xavier, M.; Oreffo, R.O.C.; Morgan, H. Skeletal stem cell isolation: A review on the state-of-the-art microfluidic label-free sorting techniques. *Biotechnol. Adv.* **2016**, *34*, 908–923. [CrossRef] [PubMed]

6. Shi, J.; Yazdi, S.; Lin, S.C.; Ding, X.; Chiang, I.K.; Sharp, K.; Huang, T.J. Three-dimensional continuous particle focusing in a microfluidic channel via standing surface acoustic waves (SSAW). *Lab Chip* **2011**, *11*, 2319–2324. [CrossRef] [PubMed]

7. Chang, C.C.; Huang, Z.X.; Yang, R.J. Three-dimensional hydrodynamic focusing in two-layer polydimethylsiloxane (PDMS) microchannels. *J. Micromech. Microeng.* **2007**, *17*, 1479–1486. [CrossRef]

8. Erb, R.M.; Martin, J.J.; Soheilian, R.; Pan, C.; Barber, J.R. Actuating soft matter with magnetic torque. *Adv. Funct. Mater.* **2016**, *26*, 3859–3880. [CrossRef]

9. Zhang, J.; Yan, S.; Yuan, D.; Zhao, Q.; Tan, S.H.; Nguyen, N.T.; Li, W. A novel viscoelastic-based ferrofluid for continuous sheathless microfluidic separation of nonmagnetic microparticles. *Lab Chip* **2016**, *16*, 3947–3956. [CrossRef]

10. Zhou, L.; Zhang, F.; Zhou, T.; Mawatari, Y. A model for estimating agglomerate sizes of non-magnetic nanoparticles in magnetic fluidized beds. *Korean J. Chem. Eng.* **2013**, *30*, 501–507. [CrossRef]

11. Pamme, N.; Manz, A. On-chip free-flow magnetophoresis: Continuous flow separation of magnetic particles and agglomerates. *Anal. Chem.* **2004**, *76*, 7250–7256. [CrossRef] [PubMed]

12. Friedman, G.; Yellen, B. Magnetic separation, manipulation and assembly of solid phase in fluids. *Curr. Opin. Colloid Interface Sci.* **2005**, *10*, 158–166. [CrossRef]

13. Nezihi Murat, K.; Spuhler, P.S.; Fabio, F.; Lim, E.J.; Vincent, P.; Emre, O.; Martel, J.M.; Nikola, K.; Kyle, S.; Pin-I, C. Microfluidic, marker-free isolation of circulating tumor cells from blood samples. *Nat. Protoc.* **2014**, *9*, 694–710. [CrossRef]

14. Zhou, Y.; Song, L.; Yu, L.; Xuan, X. Inertially focused diamagnetic particle separation in ferrofluids. *Microfluid. Nanofluid.* **2017**, *21*, 14. [CrossRef]

15. Zeng, J.; Chen, C.; Vedantam, P.; Brown, V.; Tzeng, T.R.J.; Xuan, X. Three-dimensional magnetic focusing of particles and cells in ferrofluid flow through a straight microchannel. *J. Micromech. Microeng.* **2012**, *22*, 105018. [CrossRef]

16. Zhu, J.; Liang, L.; Xuan, X. On-chip manipulation of nonmagnetic particles in paramagnetic solutions using embedded permanent magnets. *Microfluid. Nanofluid.* **2012**, *12*, 65–73. [CrossRef]

17. Sharifi, A.; Motlagh, S.Y.; Badfar, H. Ferro hydro dynamic analysis of heat transfer and biomagnetic fluid flow in channel under the effect of two inclined permanent magnets. *J. Magn. Magn. Mater.* **2019**, *472*, 115–122. [CrossRef]

18. Hennenberg, M.; Weyssow, B.; Slavtchev, S.; Desaive, T.; Scheid, B. Steady flows of a laterally heated ferrofluid layer: Influence of inclined strong magnetic field and gravity level. *Phys. Fluids* **2006**, *18*, 093602. [CrossRef]

19. Bég, O.A.; Sim, L.; Zueco, J.; Bhargava, R. Numerical study of magnetohydrodynamic viscous plasma flow in rotating porous media with Hall currents and inclined magnetic field influence. *Commun. Nonlinear Sci.* **2010**, *15*, 345–359. [CrossRef]

20. Nguyen, N.T. Micro-magnetofluidics: Interactions between magnetism and fluid flow on the microscale. *Microfluid. Nanofluid.* **2012**, *12*, 1–16. [CrossRef]

21. Rosensweig, R.E. *Ferrohydrodynamics*; Dover Publications: New York, NY, USA, 2014; pp. 54–61, 74–84, ISBN 978-0-486-67834-4.

22. Wirix-Speetjens, R.; Fyen, W.; Xu, K.; Boeck, J.D.; Borghs, G. A force study of on-chip magnetic particle transport based on tapered conductors. *IEEE Trans. Magn.* **2005**, *41*, 4128–4133. [CrossRef]

23. Kim, M.M.; Zydney, A.L. Effect of electrostatic, hydrodynamic, and Brownian forces on particle trajectories and sieving in normal flow filtration. *J. Colloid Interface Sci.* **2004**, *269*, 425–431. [CrossRef] [PubMed]

24. Gerber, R.; Takayasu, M.; Friedlaender, F.J. Generalization of HGMS theory: The capture of ultra-fine particles. *IEEE Trans. Magn.* **1983**, *19*, 2115–2117. [CrossRef]

25. Duffy, C.D.; McDonald, J.C.; Schueller, O.J.; Whitesides, G.M. Rapid prototyping of microfluidic systems in poly (dimethylsiloxane). *Anal. Chem.* **1998**, *70*, 4974–4984. [CrossRef] [PubMed]

26. Furlani, E.P.; Sahoo, Y. Analytical model for the field and force in a magnetophoretic microsystem. *J. Phys. D Appl. Phys.* **2006**, *39*, 1724–1732. [CrossRef]

27. Brody, J.P.; Yager, P.; Goldstein, R.E.; Austin, R.H. Biotechnology at low Reynolds numbers. *Biophys. J.* **1996**, *71*, 3430–3441. [CrossRef]

28. Naoki, I.; Kazuo, H.; Ryutaro, M. Interface motion of capillary-driven flow in rectangular microchannel. *J. Colloid Interface Sci.* **2004**, *280*, 155–164. [CrossRef]
29. Zhu, T.; Lichlyter, D.J.; Haidekker, M.A.; Mao, L. Analytical model of microfluidic transport of non-magnetic particles in ferrofluids under the influence of a permanent magnet. *Microfluid. Nanofluid.* **2011**, *10*, 1233–1245. [CrossRef]

Article

An Optically Induced Dielectrophoresis (ODEP)-Based Microfluidic System for the Isolation of High-Purity CD45^{neg}/EpCAM^{neg} Cells from the Blood Samples of Cancer Patients—Demonstration and Initial Exploration of the Clinical Significance of These Cells

Chia-Jung Liao [1,†], Chia-Hsun Hsieh [2,†], Tzu-Keng Chiu [3], Yu-Xian Zhu [1], Hung-Ming Wang [2], Feng-Chun Hung [1], Wen-Pin Chou [1] and Min-Hsien Wu [1,2,4,*]

[1] Graduate Institute of Biomedical Engineering, Chang Gung University, Taoyuan City 33302, Taiwan; l329735@ms49.hinet.net (C.-J.L.); maple7530@gmail.com (Y.-X.Z.); fjiun@hotmail.com (F.-C.H.); d94522010@ntu.edu.tw (W.-P.C.)

[2] Division of Haematology/Oncology, Department of Internal Medicine, Chang Gung Memorial Hospital (Linko), Taoyuan City 33302, Taiwan; wisdom5000@gmail.com (C.-H.H.); whm526@adm.cgmh.org.tw (H.-M.W.)

[3] Department of Chemical and Materials Engineering, Chang Gung University, Taoyuan City 33302, Taiwan; b74225@hotmail.com

[4] Department of Chemical Engineering, Ming Chi University of Technology, New Taipei City 24301, Taiwan

[*] Correspondence: mhwu@mail.cgu.edu.tw; Tel.: +886-3-2118800 (ext. 3599); Fax: +886-3-2118668

[†] The authors contributed equally this work.

Received: 14 October 2018; Accepted: 30 October 2018; Published: 31 October 2018

Abstract: Circulating tumour cells (CTCs) in blood circulation play an important role in cancer metastasis. CTCs are generally defined as the cells in circulating blood expressing the surface antigen EpCAM (epithelial cell adhesion molecule). Nevertheless, CTCs with a highly metastatic nature might undergo an epithelial-to-mesenchymal transition (EMT), after which their EpCAM expression is downregulated. In current CTC-related studies, however, these clinically important CTCs with high relevance to cancer metastasis could be missed due to the use of the conventional CTC isolation methodologies. To precisely explore the clinical significance of these cells (i.e., CD45^{neg}/EpCAM^{neg} cells), the high-purity isolation of these cells from blood samples is required. To achieve this isolation, the integration of fluorescence microscopic imaging and optically induced dielectrophoresis (ODEP)-based cell manipulation in a microfluidic system was proposed. In this study, an ODEP microfluidic system was developed. The optimal ODEP operating conditions and the performance of live CD45^{neg}/EpCAM^{neg} cell isolation were evaluated. The results demonstrated that the proposed system was capable of isolating live CD45^{neg}/EpCAM^{neg} cells with a purity as high as 100%, which is greater than the purity attainable using the existing techniques for similar tasks. As a demonstration case, the cancer-related gene expression of CD45^{neg}/EpCAM^{neg} cells isolated from the blood samples of healthy donors and cancer patients was successfully compared. The initial results indicate that the CD45^{neg}/EpCAM^{neg} nucleated cell population in the blood samples of cancer patients might contain cancer-related cells, particularly EMT-transformed CTCs, as suggested by the high detection rate of vimentin gene expression. Overall, this study presents an ODEP microfluidic system capable of simply and effectively isolating a specific, rare cell species from a cell mixture.

Keywords: microfluidic systems; optically induced dielectrophoresis (ODEP); cell isolation; circulating tumour cells (CTCs); cancer metastasis

---

1. Introduction

Cancer metastasis is the main cause of cancer-induced death [1]. Circulating tumour cells (CTCs), which have been documented since 1869, are cells that escape from the primary tumour site into the contiguous vasculature and subsequently exist in the blood circulation [2]. Growing evidence has revealed that the presence of CTCs in the blood circulation is linked to cancer metastasis or relapse [1]. Therefore, CTC studies have great potential for understanding the mechanisms behind cancer metastasis. This information could greatly stimulate scientists to develop new therapeutic strategies for cancer treatments. In terms of clinical applications, moreover, it has been reported that the analysis of CTCs, regarded as a liquid tumour biopsy, can be utilized as a diagnostic or prognostic tool [3] for monitoring cancer metastasis or the therapeutic response [4,5] and for guiding individualized treatment [6]. To achieve these goals, it is necessary to isolate CTCs from a blood sample at high purity to possibly avoid the analytical interferences caused by the surrounding blood cells (mainly leucocytes) [7].

Nevertheless, CTCs are very rare in a blood sample, with an approximate concentration of 1 CTC per 10^5–10^7 blood mononuclear cells [8]. This phenomenon makes CTCs difficult to isolate and purify, particularly in a high-purity manner. Leveraging the recent advances in cell isolation techniques, a wide variety of CTC isolation and purification methods have been presented and can be broadly classified as physical and biochemical methods [9]. Overall, the physical-based methods [8,10] (mainly filtration [11–13]) for CTC isolation are easy to perform and do not require labelling of the harvested cells, but the cell purity is less than the purity achieved by biochemical protocols. In the biochemical-based schemes, the techniques of immunomagnetic cell separation are predominantly adopted for CTC isolation and purification [14,15]. In these methods, magnetic beads coupled to antibodies specific to a CTCs' surface antigen (mainly the epithelial cell adhesion molecule (EpCAM) and cytokeratins (CKs)) are commonly utilized to recognize and bind the CTCs [15]. The magnetically labelled CTCs are then separated from the surrounding cells via an applied magnetic field. CTC isolation based on this strategy is primarily utilized in current CTC isolation or detection systems (e.g., the CellSearch^{TM} system [16], the magnetic-activated cell sorting system [17–19], or Dynabeads^{TM} [20]). Overall, the cell purity of CTCs obtained by the abovementioned cell isolation scheme ranges from 20% to 50% [16,19,20].

Although the abovementioned CTC isolation schemes have been successfully demonstrated, leucocyte contamination of the harvested CTCs is commonly unavoidable, possibly causing problems in the subsequent CTC-related analysis, particularly gene expression assays. This problem is primarily because the expression levels of some leucocyte genes are unclear. Therefore, their presence could interfere with subsequent analytical works [7]. This fact highlights the importance of isolating high-purity (ideally 100%) CTCs for the subsequent high-precision analyses. In addition to the CTC purity problem, there are some important biological issues that should be further considered. As discussed earlier, the majority of CTC isolation or purification strategies rely primarily on the use of EpCAM or CKs for the identification of CTCs. Nevertheless, CTCs, particularly those with a highly metastatic nature, might undergo the so-called epithelial-to-mesenchymal transition (EMT) [21]. Afterward, the CTCs might reduce their expression of EpCAM and CKs [22] and become motile cells for migration to distant metastatic sites [23]. In this situation, these clinically important CTCs associated with cancer metastasis might be missed if the conventional CTC isolation methodologies are used [24]. This possibility might, to some extent, explain the fact that these EMT-transformed CTCs are less studied in the literature.

Leveraging the technical advantages of microfluidic technology, moreover, several microfluidic systems have been demonstrated for the isolation and purification of CTCs with better performance than the conventional approaches [25]. For example, the CTC-iChip [14], two-stage microfluidic chip [26], nanostructure embedded microchips [27], parallel flow micro-aperture chip [28], and herringbone chip [29] mainly utilize EpCAM or other surface antigen-specific antibodies to recognize, separate, and then isolate CTCs from the surrounding cells in the microfluidic systems.

Although the isolation and purification of CTCs with higher purity (9.2 to 70% [14,26–28]) was achieved by utilizing these microscale devices, the problem of leucocyte contamination in the treated samples has not yet been fully solved. Otherwise, these reported devices or protocols are mainly designed for the positive isolation of CTCs (i.e., EpCAM$^{pos\ (positive)}$ CTCs), ignoring the clinically important CTCs that have undergone EMT (i.e., EpCAM$^{neg\ (negative)}$ CTCs).

To address the issues mentioned above, we attempted to integrate the techniques of fluorescence microscopic observations and optically induced dielectrophoretic (ODEP) force-based cell manipulation in a microfluidic system to isolate the specific cells of interest in a high-purity manner. In the design, immunofluorescence dye staining and fluorescence microscopic observations were first used to recognize the target cells. This recognition was followed by ODEP-based cell manipulation to precisely isolate the targeted cells from the surrounding cells. In this study, an ODEP microfluidic system was designed and fabricated. Moreover, the appropriate operating conditions (e.g., the size of the circular light image for manipulating a target cell, and the ring belt width of the circular ring dark image for repelling the non-targeted surrounding cells) for precisely manipulating cells were experimentally determined. In this work, the presented ODEP microfluidic system was utilized to isolate live CD45^{neg} and EpCAM^{neg} cells from blood samples that might contain CTCs that had undergone EMT. The performance of CD45^{neg}/EpCAM^{neg} cell isolation based on the presented method was then experimentally evaluated using a cancer cell line model. To initially explore the clinical significance of these cells (i.e., CD45^{neg}/EpCAM^{neg} cells) that are generally ignored in conventional CTC studies, a small-scale clinical experiment comparing the cancer-related gene expression levels of the CD45^{neg}/EpCAM^{neg} cells isolated from the blood samples of healthy donors and head-and-neck cancer patients was carried out.

The results revealed that the optimal diameter ratio of the target cell manipulated to the circular light image used for such cell manipulation was approximately 50% within the explored experimental conditions. In addition, the optimal ring belt width of the circular ring dark image for effective repellence of the surrounding leucocytes was 60 μm. In terms of the CD45^{neg}/EpCAM^{neg} cell isolation performance, the tested cancer cell line model demonstrated that the proposed method was able to isolate CD45^{neg}/EpCAM^{neg} cells without the contamination of leucocytes, the most abundant cell population in the treated cell samples. This result was further confirmed by the following clinical sample tests showing that the purity of the CD45^{neg}/EpCAM^{neg} cells isolated from the blood samples of cancer patients was as high as 100%, which is currently impossible when utilizing existing techniques for similar tasks. As a demonstration case, a comparison of the cancer-related gene expression levels of the CD45^{neg}/EpCAM^{neg} cells isolated from the blood samples of healthy donors and cancer patients was successfully performed. The initial results might indicate that the CD45^{neg}/EpCAM^{neg} nucleated cell population in the blood samples of cancer patients might contain cancer-related cells, particularly EMT transformed CTCs, as suggested by the high detection rate of vimentin gene expression. Overall, this study has presented an ODEP microfluidic system capable of simply and effectively isolating a specific rare cell species from a cell mixture.

2. Materials and Methods

2.1. Design of the ODEP Microfluidic System for High-Purity Isolation of CD45^{neg}/EpCAM^{neg} Cells

An ODEP microfluidic system was proposed to separate, and then isolate the CD45^{neg}/EpCAM^{neg} cells from the surrounding leucocytes in a batch-wise manner utilizing the techniques of fluorescence microscopic observation and ODEP force-based cell manipulation in a microfluidic system. The layout of the ODEP microfluidic system is illustrated in Figure 1a. In the study, a T-shaped microchannel was designed, in which the main microchannel (L: 25 mm, W: 1000 μm, H: 60 μm) was used for transportation of the cell suspension sample, and the side microchannel (L: 15 mm, W: 400 μm, H: 60 μm) was designed for the collection of the isolated cells. In the design, the fluorescence microscopic observation and the ODEP-based cell manipulations for the identification, and isolation of

CD45^{neg}/EpCAM^{neg} cells, respectively, were performed at the junction area (L: 1400 μm, W: 1000 μm, H: 60 μm) of the T-shaped microchannel, defined as the cell isolation zone. In the microfluidic system (Figure 1a), three through-holes (D: 1 mm) for tubing connections were designed, one each for loading and collecting the fresh and waste cell suspension samples, respectively, and the other one for harvesting the isolated cells. The assembly of the ODEP microfluidic system is schematically shown in Figure 1b. Briefly, the ODEP microfluidic system was composed of a top fabricated polydimethylsiloxane (PDMS) substrate (Layer A), an indium-tin-oxide (ITO) glass substrate (Layer B), double-sided adhesive tape with microfabricated microchannels (Layer C, thickness: 60 μm), and a bottom ITO glass substrate coated with a layer of photoconductive material (encompassing a 20-nm-thick n-type hydrogenated amorphous silicon layer and a 1-μm-thick hydrogenated amorphous silicon layer) (Layer D). In terms of structure, the three through-holes were located in Layer A and Layer B and were connected directly to the microchannels in Layer C.

2.2. Microfabrication and Experimental Setup

The overall fabrication process was based on a computer-numerical-controlled (CNC) machining, a metal mould-punching fabrication process, a PDMS replica moulding, a thin-film technology using sputtering and plasma-enhanced chemical vapour deposition (PECVD), and a plasma oxidation-aided bonding process [30]. Briefly, the three components of the top PDMS layer (Layer A) (Figure 1b) were fabricated by the combination of CNC machining and PDMS replica moulding, as described previously [30]. For the preparation of the ITO glass substrate (Layer B) (Figure 1b), the three through-holes were mechanically drilled in an ITO glass (15 Ω, 0.7 mm; Ritek, Taiwan) using a driller (rotational speed: 26,000 rpm). For Layer C, a custom-made metal mould was used to create the hollow structure of the T-shaped microchannels in double-sided adhesive tape (9009, 3 M, Taiwan) through a manually punching operation. For the bottom substrate (Layer D) (Figure 1b), a 70-nm-thick ITO layer was first sputtered onto a cleaned dummy glass and then thermally annealed (240 °C, 60 min). A 20-nm-thick, n-type hydrogenated amorphous silicon layer was then sputtered onto the ITO layer to improve the adhesion between the fabricated ITO glass and the amorphous silicon layer to be deposited in the subsequent process. Next, a 1-μm-thick amorphous silicon layer was deposited onto the treated ITO glass through a PECVD process [30].

In the subsequent assembly process, the three through-hole component (Layer A) was bonded with Layer B via O$_2$ plasma surface treatment. This step was followed by assembling with Layer D with the fabricated double-sided adhesive tape (Layer C) (Figure 1b). In operations, the loaded cell suspension sample was transported in the main microchannel using a suction-type syringe pump (Syringe pump 1; Figure 1c). To achieve the ODEP force-based cell manipulation [30], a function generator was used to apply an alternating current (AC) voltage between the two ITO glass layers (Layers B and D; Figure 1b) of the proposed system. A commercial digital projector (PLC-XU350, SANYO, Osaka, Japan) coupled to a computer (Computer 1; Figure 1c) was used to display light images onto the photoconductive material (Layer D) to generate an ODEP force on the manipulated cells. In addition, a CCD-equipped fluorescence microscope (Zoom 160, Optem, Fairport, NY, USA) was utilized to observe the manipulation of cells in the proposed system. The overall experimental setup is schematically illustrated in Figure 1c (a photograph of the experimental setup is provided as a Supplementary Figure: Figure S1).

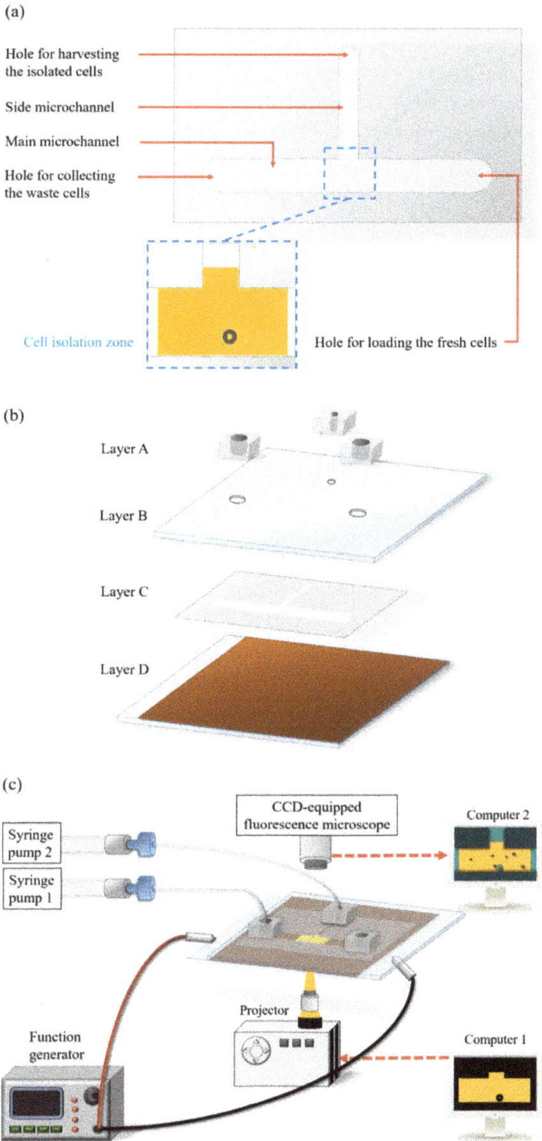

Figure 1. Schematic illustration of the (**a**) top-view layout and (**b**) assembly of the optically induced dielectrophoresis (ODEP) microfluidic system (Layer A: fabricated polydimethylsiloxane (PDMS) components; Layer B: indium-tin-oxide (ITO) glass; Layer C: double-sided adhesive tape with microfabricated microchannels; Layer D: ITO glass substrate coated with a layer of photoconductive material), and the (**c**) overall experimental setup.

2.3. The Working Scheme for the Isolation and Purification of CD45^{neg}/EpCAM^{neg} Cells

In this work, the combination of fluorescence microscopic observation and ODEP-based cell manipulation was used to identify, target, separate, and finally isolate the desired CD45^{neg}/EpCAM^{neg} cells from the surrounding leucocytes in the presented ODEP microfluidic system. The overall working procedures are illustrated in Figure 2. Briefly, all cells in a treated cell suspension were first stained with different fluorescent dyes to identify EpCAM surface marker-positive CTCs (green colour dot images), CD45 surface marker-positive leucocytes (green colour dot images), calcein AM-positive live cells (orange colour dot images), and Hoechst-positive nucleated cells (blue colour dot images). The treated cell suspension sample was then loaded in the ODEP microfluidic system, and flowed through the main microchannel (Figure 2a). During this operation, fluorescence microscopic observation was simultaneously carried out at the defined cell isolation zone to detect the desired live $CD45^{neg}$/$EpCAM^{neg}$ cells (i.e., the orange colour-only dot images). Briefly, the optical filter in the fluorescence microscope was set to observe the green and orange colour dot images. Once an orange colour-only dot image was observed, the cell suspension flow was stopped (Figure 2b). In the following procedures for the identification and positioning of the desired cells, the optical filters were switched to first observe the green colour-only dot images (i.e., the leucocytes or $EpCAM^{pos}$ CTCs) (Figure 2c) and then the blue colour-only dot images (i.e., all nucleated cells) (Figure 2d). After the contradistinction, the desired cell (i.e., the live $EpCAM^{neg}$, $CD45^{neg}$ and $Hoechst^{pos}$ cells) in the cell suspension was then targeted under light field microscopy imaging (Figure 2e). This step was followed by performing ODEP-based cell manipulation to isolate the cell targeted from the surrounding cells (Figure 2f–i). Briefly, a large-area light was used to illuminate the cell isolation zone to attract and then anchor the cells within the light-projected region, as illustrated in Figure 2f. Meanwhile, a circular ring dark image (OD: 40 μm; ID: 20 μm) was first utilized to enclose the targeted cell and soon followed by expansion of the width of the ring belt (OD: 80 μm; ID: 20 μm), as shown in Figure 2f,g. In this work, the circular ring dark image, functioning as a protective ring belt, was designed to prevent the unwanted adhesion of the surrounding cells to the manipulated cells during the operation process. The circular ring dark image around and the circular light image on the targeted cell were then moved to deliver the enclosed cell to the side microchannel, as illustrated in Figure 2g–i. After repeating the above processes, the cells of interest (i.e., the live $CD45^{neg}$/$EpCAM^{neg}$ cells) could be isolated, purified, and finally collected in the side microchannel for the subsequent collection using a suction-type syringe pump (i.e., Syringe pump 2; Figure 1c).

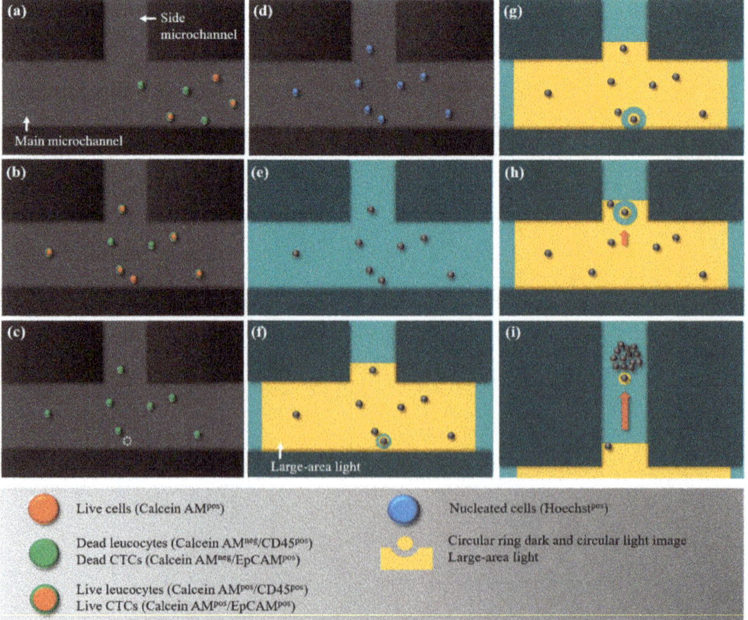

Figure 2. Schematic illustration of the overall working procedures for the isolation of $EpCAM^{neg}$ and

CD45^{neg} cells: (**a**) Fluorescence microscopic observation was performed at the defined cell isolation zone to detect the live EpCAM^{neg} and CD45^{neg} cells (i.e., the orange colour-only dot images) in a dynamic cell suspension flow; (**b**) the cell suspension flow was temporarily suspended when a live EpCAM^{neg} and CD45^{neg} cell (pointed out by the arrow marker) was observed in the cell isolation zone; (**c**) the optical filters in the fluorescence microscope were then switched to observe the green colour-only dot images (i.e., the leucocytes or EpCAM^{pos} CTCs); and (**d**) then the blue colour-only dot images (i.e., the all nucleated cells); (**e**) after the contradistinction, the desired cell (i.e., the live EpCAM^{neg}, CD45^{neg} and Hoechst^{pos} cells) in the cell suspension was then targeted under light field microscopy (pointed out by the arrow marker); and (**f**) a large-area light was used to illuminate the cell isolation zone to anchor the cells within the light-projected region. Additionally, a circular ring dark image was utilized to first enclose the targeted cell and was (**g**) soon followed by expansion of the width of the ring belt. (**h**,**i**) The circular ring dark image around and the circular light image on the targeted cell were then moved to deliver the enclosed cell to the side microchannel for the subsequent collection.

2.4. Working Mechanism and the Optimal Operating Conditions for the ODEP Force-Based Cell Manipulation in This Work

The ODEP-based microparticle manipulation first presented in 2005 [31] has been well described elsewhere [7,30,32]. Briefly, an AC electrical voltage is first applied between the two ITO glasses (e.g., Layer B and D; Figure 1b) to produce a uniform electric field between the two glass substrates. Under this circumstance, the electrically neutral microparticles (e.g., biological cells) within such an electric field become electrically polarized. The projection of light onto the photoconductive material on the bottom ITO glass (e.g., Layer D; Figure 1b) can lead to a decrease in the electrical voltage across the liquid layer within the light-illuminated region. This can thus generate a non-uniform electric field within the ODEP system. In the ODEP force-based microparticle manipulation, the interaction between the non-uniform electric field and electrically polarized particles is used to manipulate the microparticles [30]. In practical applications, one can simply control the movements of a light image to manipulate microparticles in a manageable manner.

In the proposed ODEP microfluidic system, the ODEP-based cell manipulation was utilized to separate and then isolate CD45^{neg}/EpCAM^{neg} cells from the surrounding cells, as schematically illustrated in Figure 2. In this work, the optimal operating conditions (e.g., the size of the circular light image for attracting and dragging the enclosed target cell and the ring belt width of the circular ring dark image for repelling the non-targeted surrounding cells, as illustrated in Figure 2f–i) were experimentally evaluated. In this study, the ODEP force acting on a cell was evaluated based on a method described previously [30]. Briefly, the ODEP manipulation force, which is the net force between the ODEP force and the friction force, generated on the manipulated cell was experimentally evaluated in this work. In a steady state, the ODEP manipulation force of a cell is balanced by the viscous drag of the fluid. Therefore, the hydrodynamic drag force of a moving cell is normally used to evaluate the net ODEP manipulation force of a cell according to Stokes' law (Equation (1)) [33] for describing the drag force (F) exerted on a spherical particle in a continuous flow condition.

$$F = 6\pi r \eta v \tag{1}$$

where r, η, and v denote the radius of the cell, the viscosity of the fluid, and the terminal velocity of the cell, respectively [30]. According to Stokes' law, therefore, the ODEP manipulation force can then be experimentally evaluated through the measurement of the maximum velocity of a moving light image that can manipulate a cell, as discussed previously [30].

Moreover, the ODEP force generated on a cell can be theoretically expressed by Equation (2), which was also used to describe the dielectrophoresis (DEP) force [34]:

$$F_{DEP} = 2\pi r^3 \varepsilon_m \mathrm{Re}[K(\omega)]\nabla E^2 \tag{2}$$

where r, ε_m, $K(\omega)$, and ∇E^2 denote the cell radius, the permittivity of the solution surrounding the cells, the Clausius–Mossotti factor (associated with the frequency of the electric field, the conductivity of the medium, the internal conductivity of the cells, and the membrane capacitance of the cells [35]), and the gradient of the electric field squared, respectively [36]. In terms of ODEP operating conditions, overall, the ODEP force generated on a specific cell is influenced by the magnitude and frequency of the applied electrical voltage under a given solution condition. Therefore, the electrical properties used in this work were first measured to determine the conditions under which no cell aggregation phenomenon occurs. In addition to the electrical conditions for ODEP operations, the size of the light image is also reported to influence the ODEP-based microparticle manipulation [37]. In this work, experimental evaluations were performed to determine the quantitative relationship between the diameter of the circular light image (20, 30, 40 μm) and the maximum velocity (μm s^{-1}) of the moving light image that can manipulate (i.e., attract and drag) PC-3 and SW620 cancer cells. Similar work was also carried out to evaluate the effect of the width of the dark image bar (20, 40, 60, 80 μm) on the manipulation (i.e., repulse and push) of leucocytes. After these evaluations, the optimal conditions for the light or dark image were determined for the ODEP-based cell manipulation as described in Figure 2.

2.5. Evaluation of the CD45^{neg}/EpCAM^{neg} Cell Isolation Performance-Cancer Cell Line Model

Experiments were conducted to evaluate the performance of the presented ODEP microfluidic system for CD45^{neg}/EpCAM^{neg} cell isolation. This study was approved by the Institutional Review Board of the Chang Gung Memorial Hospital. Informed consent was obtained from all blood sample donors (Approval ID: 201601081B0). All the methods were conducted in accordance with the relevant guidelines for clinical tests. Briefly, blood samples (4 mL each) were obtained from 3 healthy donors. The erythrocytes in the blood samples were first lysed using an erythrocyte lysis buffer [38]. The treated sample was further processed to deplete (CD45^{pos}) leucocytes using the EasySep Human CD45 Depletion Kit (StemCell Technologies, Vancouver, BC, Canada). After erythrocyte and leucocyte depletion, the cells remaining in the treated samples were stained with fluorescent dyes for identifying EpCAM surface marker-positive CTCs (green colour dot images) (Alexa Fluor 488-labelled EpCAM antibody, Cell Signalling Technology, Danvers, MA, USA), CD45 surface marker-positive leucocytes (green colour dot images) (FITC-labelled CD45 antibody, eBioscience, San Diego, CA, USA), calcein AM-positive live cells (orange colour dot images) (Thermo Fisher Scientific, Waltham, MA, USA), and Hoechst-positive nucleated cells (blue colour dot images) (Thermo Fisher Scientific). In this work, SW620 (cancer cell line) cells were spiked into the abovementioned treated cell samples to mimic the existence of CD45^{neg}/EpCAM^{neg} cells in the cell samples. According to the previous report [39], there are about 10^2 to 10^4 CD45neg/EpCAMneg cells/mL in the blood samples of head-and-neck cancer patients. Based on this result, therefore, we spiked 2000 cells to 4 mL of blood sample from healthy donor (i.e., 500 CD45^{neg}/EpCAM^{neg} cells/mL). In general, there will be about 2×10^7 cells remained in the sample after 4 mL of blood sample is treated with blood cell depletion processes. Therefore, the percentage of the CD45^{neg}/EpCAM^{neg} cells spiked in a treated blood sample was about 0.01% ($2000/2 \times 10^7 \times 100\%$). Briefly, 2000 SW620 cancer cells stained with the abovementioned fluorescent dyes except for the EpCAM surface marker dye were spiked into the treated cell samples. The treated cell samples were then prepared in 100 μL of 300 mM sucrose solution (solution conductivity: 3.4 μS cm^{-1}). This step was soon followed by the ODEP-based CD45^{neg}/EpCAM^{neg} cell isolation described in Figure 2. The gene expression of the harvested CD45^{neg}/EpCAM^{neg} cells was then analysed using the real-time polymerase chain reaction (PCR) technique. Briefly, the total RNA of the cells was extracted using the PicoPure RNA isolation Kit (Thermo Fisher Scientific) and the cDNA was then synthesized via the SuperScript IV First-Strand Synthesis System (Thermo Fisher Scientific). This step was followed by cDNA pre-amplification using TaqMan PreAmp Master Mix (Thermo Fisher Scientific) to increase the quantity of the target genes. Afterward, TaqMan-based detection was carried to determine the gene expression levels of the isolated cells. In this work, the TaqMan assays for each gene were purchased from Thermo Fisher Scientific and were performed according to the manufacturer's instructions. The β-2-microglobulin serves as the internal control gene in this study.

2.6. Comparison of the Cancer-Related Gene Expression of the CD45^{neg}/EpCAM^{neg} Cell Populations Isolated from the Blood Samples of Healthy Donors and Head-and-Neck Cancer Patients

To initially explore the clinical significance of the CD45^{neg}/EpCAM^{neg} cell populations in the blood samples of cancer patients, the following small-scale clinical test was carried out. Briefly, blood samples (4 mL each) were obtained from patients diagnosed with head-and-neck cancer ($n = 8$) and from healthy blood donors ($n = 5$). The blood samples were then processed using the protocol described earlier to isolate the CD45^{neg}/EpCAM^{neg} cell population. In this study, we only harvested about 50 CD45^{neg}/EpCAM^{neg} cells in a blood sample for the subsequent gene expression analysis. This is mainly because 25–50 cells were technically sufficient for the subsequent analysis of their gene expression. The harvested cells were then analysed to determine their cancer-related gene expression using real-time PCR as described earlier.

2.7. Statistical Analysis

Data from at least three separate experiments were analysed and presented as the mean ± the standard deviation. One-way analysis of variance (ANOVA) was used to examine the effect of the experimental conditions on the results. The Tukey honestly significant difference (HSD) post hoc test was used to compare the differences between two investigated conditions when the null hypothesis of ANOVA was rejected.

3. Results and Discussions

3.1. Characteristic Features of the Proposed ODEP-Based Microfluidic System for the Isolation and Purification of CD45^{neg}/EpCAM^{neg} Cells

In this study, the integration of fluorescence microscopic observation and ODEP force-based cell manipulation in a microfluidic system was proposed for high-purity isolation of CD45^{neg}/EpCAM^{neg} cells based on the working schemes described in Figure 2. First, the presented ODEP microfluidic system features in providing a gentler process for separating and isolating viable cells from a cell mixture than current techniques [30,40]. This technical advantage might be difficult to achieve using other microfluidic systems designed for the same purpose, in which the isolated cells might be damaged due to the high fluidic shear stress in a microfluidic system. This characteristic is found to be valuable for using the harvested viable cells for subsequent gene expression analysis. Additionally, in terms of the cell manipulation technique, a more user-friendly and flexible ODEP force-based working mechanism was adopted in this design compared to that in the other methods (e.g., techniques based on fluidic control [40], magnetic force [14], thermal control [41], or dielectrophoretic force (DEP) [42]) reported in the literature. Among the cell manipulation methods, the DEP force-based mechanism has been actively proposed for a wide variety of applications [43], mainly due to its capability of providing precise cell manipulation and control. Nevertheless, the DEP-based method requires a technically demanding, costly, and lengthy microfabrication process to create a unique metal microelectrode array on a substrate that is specific to the application. This technical shortcoming can be adequately solved by using the ODEP force-based technique, by which a specific microelectrode layout can be simply and flexibly created or modified through the manipulation of the optical patterns projected on the photoconductive material-coated ITO glass of the ODEP system (e.g., Figure 1c) that act as a virtual electrode layout [31]. In practice, one can simply utilize a commercial digital projector to display optical patterns on the ODEP system to manipulate cells in a simple, flexible, and user-friendly manner via computer-interface controls [30].

In terms of cell-related studies, the utilization of the ODEP-based mechanism in microfluidic systems has been presented for various applications, mainly cell patterning [31], cell lysis [44], cell fusion [45], cell sorting based on the cellular size [30] or electrical properties [46], and cell isolation and purification [7]. For the last example, ODEP-based methods have been presented for the isolation of bacteria [47], droplets [48], and CTCs [7]. For CTC isolation, the ODEP technique was found to be particularly promising for the isolation of rare cells in a cell mixture in a high-purity manner due to its ability for precise cell manipulation [7]. The level of cell isolation achieved by such a methodology is normally technically impossible by the conventional macro-scale devices or methods [14,16,20]. In operations, light images are commonly used to target the wanted cells, and such light images are then used to manipulate and then separate the targeted cells from the surrounding cells for cell isolation and purification purposes [7]. It was recently reported that the CTC isolation techniques based on this method can achieve a CTC cell purity as high as 100% [7], which is currently impossible using the other existing methods. During the cell manipulation process using moving light images, however, cautious operations are normally required to prevent the light images from attracting unwanted cells around the target cells [7]. Otherwise, the purity of the harvested cells might be seriously affected. To address this issue, one of the technical features of the presented method is to use a large-area light to illuminate the cell isolation zone to attract and then immobilize all cells within this zone on the surface of a substrate (Figure 2f). Afterward, a circular ring dark image, functioning as a protective ring belt, was designed to enclose the targeted cell, effectively preventing the unwanted adhesion of the surrounding cells to the manipulated cells during the target cell isolation process, as described in Figure 2f–h. This design not only provides high-purity isolation of the target cells but also makes the cell manipulation process simpler and more user-friendly.

3.2. Optimal Operating Conditions for ODEP Force-Based Cell Manipulation

In this study, ODEP-based cell manipulation was utilized to separate and then isolate the $CD45^{neg}/EpCAM^{neg}$ cells from the surrounding cells, as illustrated in Figure 2. For effective and efficient cell isolation, the optimal operating conditions were experimentally evaluated. As described earlier in Equation (2), the ODEP force generated on a specific cell is influenced by the magnitude and frequency of the electrical voltage applied under a given solution condition. In this study, the electrical condition of 8 V (3 MHz) was first evaluated experimentally, and no cell aggregation phenomenon was observed under such condition. This result could greatly facilitate the subsequent cell isolation process. To determine the optimum size of light and dark images for cell manipulation, experimental evaluations were also performed. In this work, two kinds of cancer cell lines with different sizes (microscopically estimated diameter: 20.1 ± 1.5 and 10.5 ± 0.9 μm for PC-3 and SW620 cancer cells, respectively) were used as model cells for the testing. In this study, the ODEP manipulation force acting on a cell was evaluated through measurement of the maximum velocity of a moving light (dark) image that can manipulate a cell, as discussed previously [7]. In this evaluation, the quantitative relationship between the diameter (20, 30, 40 μm) of the circular light image that was used (i.e., the light illuminating the targeted cell; Figure 2f–i) and the maximum velocity (μm s^{-1}) of the moving light image that can manipulate (i.e., attract and drag) PC-3 and SW620 cancer cells was first established. Similar work was also carried out to evaluate the effect of the dark bar image width (20, 40, 60, 80 μm) on the manipulation (i.e., repulse and push) of the surrounding cells (mainly leucocytes).

The results (Figure 3a,b) revealed that the diameter of the circular light image had a significant influence ($p < 0.01$, ANOVA) on the maximum velocity (μm s^{-1}) of the moving light image that can manipulate the SW620 and PC-3 cancer cells. This result was in line with the previous findings demonstrating that the size of a light image can play a role in the ODEP-based microparticle manipulation [37]. For the SW620 cancer cells with average diameter of 10.5 ± 0.9 μm, the measured maximum velocity that could manipulate this cell type significantly ($p < 0.01$) decreased when the largest circular light image (D: 40 μm) was used with the explored experimental conditions (Figure 3a). This finding can be explained by the fact that under an ODEP field, the electric field within a smaller light image could be more focused than that in a larger image. This feature could accordingly contribute to a higher ODEP manipulation force and thus higher maximum velocity for the light image manipulation of a cell [7]. For the PC-3 cancer cells with an average diameter of 20.1 ± 1.5 µm, conversely, the measured maximum velocity significantly ($p < 0.05$) decreased when the smallest circular light image (D: 20 µm) was used (Figure 3b). This unexpected result could be because the circular light image with the smallest size (D: 20 µm) was close to the average size (20.1 ± 1.5 µm) of the PC-3 cancer cells used in this study. This similarity could, therefore, lead to instability in the ODEP force generation, as discovered previously [7].

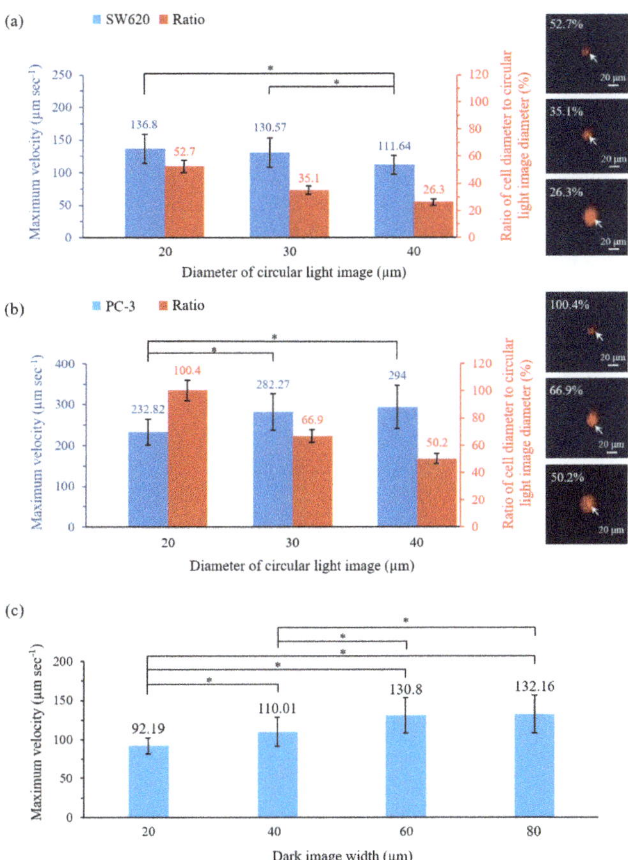

Figure 3. Quantitative relationship between the diameter of the circular light image (and the diameter ratio of the cell to the circular light image) that is used and the maximum velocity (µm s⁻¹) of the moving circular light image that can manipulate (i.e., attract and drag) (**a**) SW620 and (**b**) PC-3 cancer cells (right column: the microscopic images of various diameter ratios of cell to circular light image; the cells are pointed out by arrow markers) and (**c**) the quantitative link between the dark image width that is used and the maximum velocity (µm s⁻¹) of the moving dark image that can manipulate (i.e., repulse and push) leucocytes.

In terms of the ratio of the cell diameter to the circular light image diameter, overall, the maximum velocity of the circular light image that could manipulate a cell was higher when the ratio was in the range of 35.1–52.7% and 50.2–66.9% for the SW620 and PC-3 cancer cells, respectively (Figure 3a,b). Based on these results, the ratio was set at 50% for the following experiments. In this study, moreover, a circular ring dark image was designed to enclose the abovementioned circular light image when a cell was targeted. As described earlier, this design was to repel the surrounding cells (mainly leucocytes) and thus, to prevent the unwanted adhesion of leucocytes to the manipulated cells during the target cell isolation process (Figure 2f–i). In this study, a similar evaluation was carried out to determine the optimal ring belt width of the circular ring dark image for the manipulation (i.e., repulse and push) of the leucocyte. The results (Figure 3c) exhibited that the size of the ring belt width had a significant ($p < 0.01$; ANOVA) impact on the leucocyte manipulation, in which the maximum velocity of the dark bar image that could manipulate a leucocyte increased as the dark bar image width increased from 20 µm to 60 µm. According to the results (Figure 3c), therefore, the ring belt width of the circular ring dark image was set at 60 µm for the following works.

3.3. Performance Evaluation of $CD45^{neg}/EpCAM^{neg}$ Cell Isolation

In this study, the performance of the proposed ODEP microfluidic system for the isolation of $CD45^{neg}/EpCAM^{neg}$ cells was experimentally assessed. First, the erythrocytes and leucocytes in the blood samples of healthy donors were mostly depleted. The remaining cells in the treated samples were then stained with four kinds of fluorescent dyes, as described in Figure 2. In this work, SW620 cancer cells stained with the four fluorescence dyes except for the EpCAM antibody-based dye were spiked into the abovementioned treated cell samples. Although SW620 cancer cells normally have surface antigen expression of EpCAM in nature, these cells were not stained with the EpCAM antibody-based dye in this study, making them non-observable in the fluorescence microscopic observation for the identification of $CD45^{pos}$ or $EpCAM^{pos}$ cells, as described in Figure 2a–c. Based on this design, therefore, the SW620 cancer cells spiked in a sample would be grouped into the $CD45^{neg}/EpCAM^{neg}$ cell population in this study. This design was to mimic the existence of cancer-related cells in the $CD45^{neg}/EpCAM^{neg}$ cell population in the cell samples. Afterward, the live $CD45^{neg}/EpCAM^{neg}$ nucleated cells were isolated from the cell samples based on the procedures described in Figure 2. The recovery rate of the device was evaluated to be $81.0 \pm 0.7\%$. In this study, 50 cells were isolated and delivered to the side microchannel for the subsequent collection. Practically, about 40 cells were obtained after the subsequent collection process. The harvested live $CD45^{neg}/EpCAM^{neg}$ nucleated cells were then analysed to determine their expression of leucocyte-specific (e.g., CD45) and cancer cell-specific (e.g., EpCAM, and CK19) genes using a real-time PCR system.

The results (Figure 4a) revealed that the expression of the CD45 gene was detectable only in pure leucocytes (as a control) but not in pure SW620 cancer cells, and the $CD45^{neg}/EpCAM^{neg}$ nucleated cells harvested via the proposed cell isolation scheme. This result demonstrated that the $CD45^{neg}/EpCAM^{neg}$ nucleated cells harvested through the proposed ODEP microfluidic system were pure and not contaminated with leucocytes, the most abundant cell population in the treated cell samples. This technical advantage was found to be valuable for using the harvested cells for the subsequent high-precision gene expression analysis because the existence of the surrounding cells, such as leucocytes, could cause analytical interference [7]. Moreover, the expression of the EpCAM and CK19 genes was detected only in the harvested live $CD45^{neg}/EpCAM^{neg}$ nucleated cells but not in leucocytes (as a control) (Figure 4b,c). This result indicated that the spiked SW620 cancer cells mimicking the existence of cancer-related cells in the $CD45^{neg}/EpCAM^{neg}$ cell population were able to be successfully isolated and detected by the proposed cell isolation process, and the protocol for gene expression analysis, respectively.

In addition to the cancer cell line model described above, the performance of $CD45^{neg}/EpCAM^{neg}$ cell isolation was also evaluated using a clinical sample model. In this work, blood samples were obtained from patients diagnosed with head-and-neck cancer. The samples were first treated, and the $CD45^{neg}/EpCAM^{neg}$ cell isolation process was then performed as described in Figure 2. Figure 5 shows the real cell isolation operation process. Briefly, Figure 5a(I–III) demonstrated the process of fluorescence microscopic observation to identify the $CD45^{neg}/EpCAM^{neg}$ cells marked by the arrow; these cells were found in both the orange colour-only dot image (Figure 5a(I–II)) and the blue colour dot image (Figure 5a(III)). After the $CD45^{neg}/EpCAM^{neg}$ cell was positioned using light field microscopy imaging (Figure 5a(IV)), ODEP-based cell manipulation was carried out to isolate the cells targeted to the side microchannel (Figure 5a(V–VII)). Through repetition of the processes abovementioned, the $CD45^{neg}/EpCAM^{neg}$ cells were able to be isolated and temperately stored in the side microchannel for the subsequent collection. In this work, the purity of the live $CD45^{neg}/EpCAM^{neg}$ cells in the side microchannel was further evaluated microscopically. The results (Figure 5b) revealed that the purity of the live $CD45^{neg}/EpCAM^{neg}$ cells was as high as 100%, which is beyond the currently possible purity obtained by using other existing cell isolation techniques. In this proof-of-concept study, the proposed ODEP-based cell isolation system has not yet been automated. The operating time for processing 4 mL of blood sample was about 4 h, in which the manual fluorescent microscopic observation consumed most of the time required. As a result, the automation of the proposed system will be our future goal, which can largely improve the working throughput of the proposed ODEP-based cell isolation system.

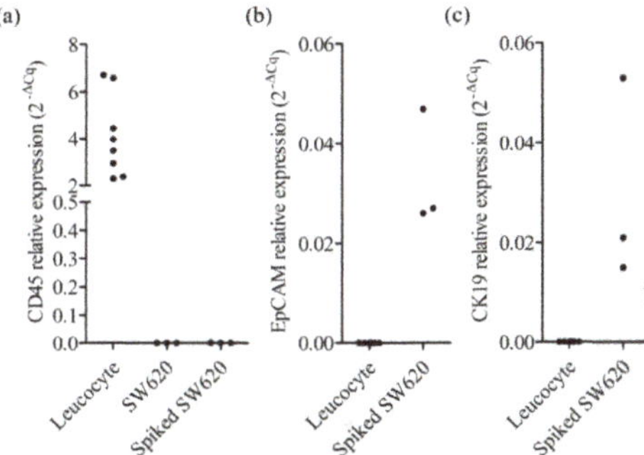

Figure 4. Analysis of the gene ((**a**) leucocyte-specific CD45 and cancer cell-specific (**b**) EpCAM and (**c**) CK19 genes) expression levels in the pure leucocytes, the SW620 cancer cells, and the cells isolated by the proposed method.

3.4. Comparison of the Cancer-Related Gene Expression of the $CD45^{neg}/EpCAM^{neg}$ Cells Isolated from the Blood Samples of Healthy Donors and Head-and-Neck Cancer Patients

In conventional CTC-related studies, the cellular proteins EpCAM and CKs are predominately used as biomarkers to identify CTCs [15]. However, growing evidence has suggested that the use of these biomarkers to identify CTCs is not sufficient due to the heterogeneous characteristics of CTCs [24]. As a result, the search for other more clinically meaningful markers that might be more relevant to cancer metastasis has become important. It is well recognized that the CTCs with a highly metastatic nature might undergo EMT [21], after which their expression of EpCAM and CKs is downregulated [22]. Therefore, these cells are generally ignored in the conventional positive selection-based CTC isolation schemes. In our recent study, moreover, it was discovered that the number of $CD45^{neg}/EpCAM^{neg}$ nucleated cells in the blood samples of cancer patients is significantly higher than that of healthy donors [49]. To initially explore the clinical significance of the $CD45^{neg}/EpCAM^{neg}$ nucleated cells, in which the EMT transformed CTCs might contain, a preliminary clinical test was carried out. In this work, the $CD45^{neg}/EpCAM^{neg}$ nucleated cells were isolated from the blood samples of patients diagnosed with head-and-neck cancer and from healthy donors based on the processes described in Figure 2. The cancer-related gene expression of the $CD45^{neg}/EpCAM^{neg}$ nucleated cells in the two sample groups was analysed and then compared. In this work, the investigated cancer-related genes were genes related to EMT (CDH1, CDH2, EpCAM, CK19, Vimentin, SNAIL1, TWIST1) [50], multiple drug resistance (MRP1, MRP2, MRP4, MRP5, MRP7) [51], and cancer stem cell (CSC) (ALDH1, NANOG, OCT4, SOX2, CD133) [52].

(a)

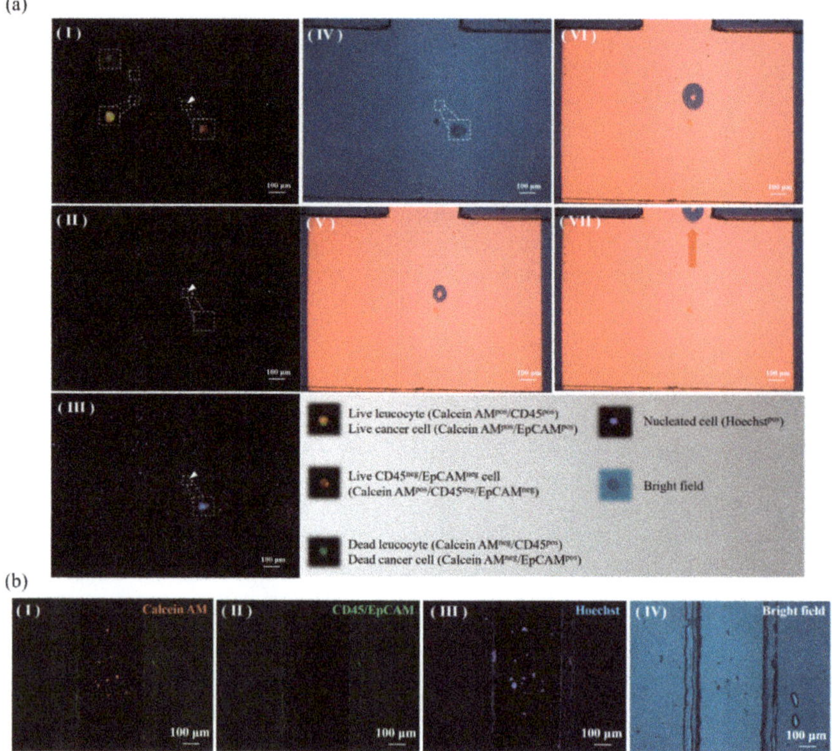

(b)

Figure 5. (**a**) Microscopic images showing the isolation of CD45^{neg}/EpCAM^{neg} cells from the blood sample of head-and-neck cancer patients: (I)–(III) the processes of the fluorescence microscopic observation to identify the CD45^{neg}/EpCAM^{neg} cells ((I)–(II) the orange colour-only dot image and (III) the blue colour dot image), (IV) the target CD45^{neg}/EpCAM^{neg} cell was positioned using light field microscopy imaging, (V)–(VII) ODEP-based cell manipulation was carried out to isolate the targeted cells to the side microchannel (a video clip is provided in the supplementary material); (**b**) microscopic observations for evaluating the purity the live CD45^{neg}/EpCAM^{neg} cells isolated and collected in the side microchannel ((I) calcein AM-positive live cells (orange colour dot images), (II) CD45 surface marker-positive leucocytes, or EpCAM surface marker-positive CTCs (green colour dot images); (III) Hoechst-positive nucleated cells (blue colour dot images); (IV) the cell image obtained using light field microscopy).

Table 1 provides a summary of the results of this comparison (the detailed experimental results are provided as a Supplementary Table: Table S1). In terms of the EMT-related gene expression (Table 1), briefly, EpCAM gene expression was not detected in the two compared sample groups because EpCAM^{pos} cells were excluded in the proposed CD45^{neg}/EpCAM^{neg} cell isolation scheme. This result further confirmed that the purity of the CD45^{neg}/EpCAM^{neg} cells isolated was high, consistent with the findings in Figure 5. Expression of the CK19 (i.e., gene for an epithelial marker) and SNAIL1 (i.e., gene for a zinc finger transcription factor that drives EMT progression [53]) genes was not detected in the healthy blood sample group but was detected in 1 of 8 cancer patients (i.e., detection date: 12.5%). Vimentin is the major protein constituent of the intermediate filament responsible for the maintenance of the cytoskeleton structure of a cell. It has been reported that vimentin is involved in tumour progression, cell migration, and tumour invasion [54]. Over-expression of vimentin in diverse tumours has been discovered previously [55,56]. In addition, vimentin expression in CTCs was also reported [57,58], and the detection of vimentin-positive CTCs has been associated with disease progression [59,60]. However, the detection of vimentin gene expression in the CD45^{neg}/EpCAM^{neg} cells has not yet been reported. In this work, vimentin gene expression was detected in 1 of 5 blood samples from the healthy donors (20%) and in 6 of 8 blood samples from the cancer patients (75%). Apart from the gene expression discussed above, the expression of other evaluated EMT-related genes, such as TWIST1, CDH1, and CDH2, was undetectable in the two sample groups. Taken together, the results described above might indicate that EMT-transformed CTCs might exist in the CD45^{neg}/EpCAM^{neg} cell population.

Table 1. Gene expression status of CD45^{neg}/EpCAM^{neg} nucleated cells isolated from the blood samples of healthy donors and head-and-neck cancer patients.

Biological Function	Gene Name	Healthy Donor	Head-and-Neck Cancer Patient
		Positive/Total	
EMT-related	EpCAM	0/5	0/8
	CK19	0/5	1/8
	Vimentin	1/5	6/8
	SNAIL1	0/5	1/8
MDR-related	MRP1	0/5	3/8
	MRP2	0/5	0/8
	MRP4	0/5	0/8
	MRP5	0/5	2/8
	MRP7	0/5	0/8
CSC-related	NANOG	1/5	2/8
	OCT4	1/5	4/8

The gene expression of multi-drug resistance proteins (MRPs) including MRP1, MRP2, MRP4, MRP5, and MRP7 was not detected in the CD45^{neg}/EpCAM^{neg} nucleated cells isolated from the blood samples of healthy donors (Table 1). In addition, the expression of these genes, except for the MRP1 and MRP5 genes, also was not detected in the cells isolated from the blood samples of cancer patients. Expression of the MRP1 and MRP5 genes was detected in 3 and 2 of 8 samples from cancer patients (i.e., detection rate: 37.5% and 25.0%, respectively), respectively. The MRPs belong to the family of adenosine triphosphate -binding cassette transporters and are thought to be responsible for transporting nutrients, metabolites or drugs through a cell [51]. In cancer, the over-expression of MRPs is associated with cancer cell resistance to anticancer drugs [51]. Therefore, monitoring the MRP gene expression of cancer cells could provide valuable information relevant to the disease status during or after therapies [51]. In this work, the MRP gene expression in the CD45^{neg}/EpCAM^{neg} nucleated cells was studied for the first time. However, the correlation between the MRP gene expression of the CD45^{neg}/EpCAM^{neg} nucleated cells and the treatment response or disease progression of cancer patients was not explored.

Cancer stem cells (CSCs) have been found in many types of cancer, including head-and-neck cancer [61]. Cancer cells with the characteristics of stem cells are reported to be associated with cancer progression and treatment resistance [62]. Therefore, the evaluation of CSC-related gene expression in CD45^{neg}/EpCAM^{neg} cells might provide information relevant to the cancer disease state. In this work, the expression of CSC-related genes (e.g., ALDH1A1, CD133, NANOG, OCT4, and SOX2) was analysed in the isolated CD45^{neg}/EpCAM^{neg} nucleated cells. The results (Table 1) exhibited that only two (i.e., NANOG and OCT4) of these genes were detected in these cells isolated from the blood samples of cancer patients (2 and 4 of 8 samples, respectively), although the expression of these two genes was also found in 1 of 5 blood samples of healthy donors. It seemed that the detection rate of NANOG and OCT4 gene expression in the cancer patient group (25%, and 50%, respectively) was slightly higher than that in the healthy blood donor group (20%). In this study, as a whole, the EMT-, MRP-, and CSC-related gene expression of the CD45^{neg}/EpCAM^{neg} nucleated cells isolated from the blood samples of cancer patients and healthy donors was reported for the first time. The overall results described above might indicate that the $CD45^{neg}/EpCAM^{neg}$ nucleated cells isolated from the blood samples of cancer patients might contain cancer-related cells (e.g., EMT-transformed CTCs due to the high detection rate of vimentin gene expression). Further clinical study on a large scale is required to confirm the results reported in this work. Additionally, a clinical study that links the correlations between the cancer-related gene expression of $CD45^{neg}/EpCAM^{neg}$ nucleated cells isolated from the blood samples of cancer patients and the cancer disease status might be worthwhile.

4. Conclusions

The current research of CTCs or cells relevant to cancer metastasis (e.g., the $CD45^{neg}/EpCAM^{neg}$ cells in this study) is normally hindered by the lack of an adequate cell isolation method capable of efficiently and effectively harvesting these rare cell species with high purity. To address this issue, this study proposed the integration of the techniques of fluorescence microscopic observations and ODEP-based cell manipulation in a microfluidic system to isolate the specific cells of interest in a high-purity manner. In the design, the immunofluorescence dye staining and fluorescence microscopic observations were first used to recognize the targeted cells. This step was followed by ODEP-based cell manipulation to precisely isolate the targeted cells from the surrounding cells. Compared with the other microfluidic systems used for the similar tasks, the proposed ODEP microfluidic system mainly provides a way to harvest cells in a high-purity, cell-friendly, and low-cost manner. Moreover, due to the specific design of light images (i.e., circular ring dark image), the presented system provides a method to effectively prevent the unwanted adhesion of the surrounding cells to the manipulated cells during the target cell isolation process. This design not only provides higher purity isolation of the target cells but also makes the cell manipulation process simpler and more user-friendly than other ODEP-based microfluidic systems used for the same purpose. In this study, the optimal size of optical images (i.e., 50% for the diameter ratio of the manipulated target cell to the circular light image used for such cell manipulation, and 60 μm for the ring belt width of the circular ring dark image) for ODEP-based cell manipulation was experimentally evaluated. In the cancer cell line model, the proposed method was proved to be able to isolate $CD45^{neg}/EpCAM^{neg}$ cells without the contamination of leucocytes, the most abundant cell population in the treated cell samples. This performance evaluation was further confirmed by the subsequent clinical sample tests demonstrating that the purity of the $CD45^{neg}/EpCAM^{neg}$ cells isolated from the blood samples of metastatic cancer patients was as high as 100%, which is currently impossible via the existing techniques for similar tasks. In the comparison of the cancer-related gene expression in the $CD45^{neg}/EpCAM^{neg}$ cells isolated from the blood samples of healthy donors and cancer patients, the initial results indicate that the $CD45^{neg}/EpCAM^{neg}$ nucleated cell population might contain cancer-related cells, particularly EMT-transformed CTCs, as suggested by the high detection rate of vimentin gene expression. Further clinical study on a large scale is required to confirm the results reported in this work.

Supplementary Materials: The following are available online at http://www.mdpi.com/2072-666X/9/11/563/s1. Figure S1: Photograph of the overall experimental setup. Table S1: Gene expression status of $CD45^{neg}/EpCAM^{neg}$ nucleated cells isolated from the blood samples of healthy donors and head-and-neck cancer patients. Video S1: ODEP-based cell manipulation.

Author Contributions: C.-J.L., C.-H.H., and T.-K.C. contributed to conception and design, data analysis, and interpretation. Y.-X.Z. and F.-C.H. contributed to data acquisition and data analysis. H.-M.W. and W.-P.C. contributed to final approval of the version to be published. M.-H.W. contributed to agreement to be accountable for all aspects of the work in ensuring that questions related to the accuracy or integrity of the work are appropriately investigated and resolved.

Acknowledgments: This work was sponsored by the Ministry of Science and Technology, (MOST 104-2628-E-182-002-MY3 and 105-2221-E-182-028-MY3) and Chang Gung Memorial Hospital (CMRPG3G0591-3, CMRPG3E1633, CMRPD2E0011-13, CMRPD2G0061-62, and CMRPD2H0121).

Conflicts of Interest: The authors declare no conflict of interest.

References

1. Mehlen, P.; Puisieux, A. Metastasis: A question of life or death. *Nat. Rev. Cancer* **2006**, *6*, 449–458. [CrossRef] [PubMed]
2. Ashworth, T. A case of cancer in which cells similar to those in the tumours were seen in the blood after death. *Aust. Med. J.* **1869**, *14*, 146.
3. Zhang, Y.; Lv, Y.; Niu, Y.; Su, H.; Feng, A. Role of Circulating Tumor Cell (CTC) Monitoring in Evaluating Prognosis of Triple-Negative Breast Cancer Patients in China. *Med. Sci. Monit.* **2017**, *23*, 3071–3079. [CrossRef] [PubMed]
4. Dago, A.E.; Stepansky, A.; Carlsson, A.; Luttgen, M.; Kendall, J.; Baslan, T.; Kolatkar, A.; Wigler, M.; Bethel, K.; Gross, M.E.; et al. Rapid phenotypic and genomic change in response to therapeutic pressure in prostate cancer inferred by high content analysis of single circulating tumor cells. *PLoS ONE* **2014**, *9*, e101777. [CrossRef] [PubMed]
5. Chen, C.L.; Mahalingam, D.; Osmulski, P.; Jadhav, R.R.; Wang, C.M.; Leach, R.J.; Chang, T.C.; Weitman, S.D.; Kumar, A.P.; Sun, L.; et al. Single-cell analysis of circulating tumor cells identifies cumulative expression patterns of EMT-related genes in metastatic prostate cancer. *Prostate* **2013**, *73*, 813–826. [CrossRef] [PubMed]
6. Van Keymeulen, A.; Lee, M.Y.; Ousset, M.; Brohee, S.; Rorive, S.; Giraddi, R.R.; Wuidart, A.; Bouvencourt, G.; Dubois, C.; Salmon, I.; et al. Reactivation of multipotency by oncogenic PIK3CA induces breast tumour heterogeneity. *Nature* **2015**, *525*, 119–123. [CrossRef] [PubMed]
7. Chiu, T.K.; Chou, W.P.; Huang, S.B.; Wang, H.M.; Lin, Y.C.; Hsieh, C.H.; Wu, M.H. Application of optically-induced-dielectrophoresis in microfluidic system for purification of circulating tumour cells for gene expression analysis- Cancer cell line model. *Sci. Rep.* **2016**, *6*, 32851. [CrossRef] [PubMed]
8. Joosse, S.A.; Gorges, T.M.; Pantel, K. Biology, detection, and clinical implications of circulating tumor cells. *EMBO Mol. Med.* **2015**, *7*, 1–11. [CrossRef] [PubMed]
9. Sun, Y.; Haglund, T.A.; Rogers, A.J.; Ghanim, A.F.; Sethu, P. Review: Microfluidics technologies for blood-based cancer liquid biopsies. *Anal. Chim. Acta* **2018**, *1012*, 10–29. [CrossRef] [PubMed]
10. Yousuff, C.; Ho, E.; Hussain, K.I.; Hamid, N. Microfluidic Platform for Cell Isolation and Manipulation Based on Cell Properties. *Micromachines* **2017**, *8*, 15. [CrossRef]
11. Hvichia, G.E.; Parveen, Z.; Wagner, C.; Janning, M.; Quidde, J.; Stein, A.; Muller, V.; Loges, S.; Neves, R.P.; Stoecklein, N.H.; et al. A novel microfluidic platform for size and deformability based separation and the subsequent molecular characterization of viable circulating tumor cells. *Int. J. Cancer* **2016**, *138*, 2894–2904. [CrossRef] [PubMed]
12. Fan, X.; Jia, C.; Yang, J.; Li, G.; Mao, H.; Jin, Q.; Zhao, J. A microfluidic chip integrated with a high-density PDMS-based microfiltration membrane for rapid isolation and detection of circulating tumor cells. *Biosens. Bioelectron.* **2015**, *71*, 380–386. [CrossRef] [PubMed]
13. Vona, G.; Sabile, A.; Louha, M.; Sitruk, V.; Romana, S.; Schutze, K.; Capron, F.; Franco, D.; Pazzagli, M.; Vekemans, M.; et al. Isolation by size of epithelial tumor cells: A new method for the immunomorphological and molecular characterization of circulatingtumor cells. *Am. J. Pathol.* **2000**, *156*, 57–63. [CrossRef]
14. Ozkumur, E.; Shah, A.M.; Ciciliano, J.C.; Emmink, B.L.; Miyamoto, D.T.; Brachtel, E.; Yu, M.; Chen, P.I.; Morgan, B.; Trautwein, J.; et al. Inertial focusing for tumor antigen-dependent and -independent sorting of rare circulating tumor cells. *Sci. Transl. Med.* **2013**, *5*, 179ra147. [CrossRef] [PubMed]
15. Deng, G.; Herrler, M.; Burgess, D.; Manna, E.; Krag, D.; Burke, J.F. Enrichment with anti-cytokeratin alone or combined with anti-EpCAM antibodies significantly increases the sensitivity for circulating tumor cell detection in metastatic breast cancer patients. *Breast Cancer Res.* **2008**, *10*, R69. [CrossRef] [PubMed]
16. Riethdorf, S.; Fritsche, H.; Muller, V.; Rau, T.; Schindlbeck, C.; Rack, B.; Janni, W.; Coith, C.; Beck, K.; Janicke, F.; et al. Detection of circulating tumor cells in peripheral blood of patients with metastatic breast cancer: A validation study of the CellSearch system. *Clin. Cancer Res.* **2007**, *13*, 920–928. [CrossRef] [PubMed]
17. Mostert, B.; Sleijfer, S.; Foekens, J.A.; Gratama, J.W. Circulating tumor cells (CTCs): Detection methods and their clinical relevance in breast cancer. *Cancer Treat. Rev.* **2009**, *35*, 463–474. [CrossRef] [PubMed]
18. Martin, V.M.; Siewert, C.; Scharl, A.; Harms, T.; Heinze, R.; Ohl, S.; Radbruch, A.; Miltenyi, S.; Schmitz, J. Immunomagnetic enrichment of disseminated epithelial tumor cells from peripheral blood by MACS. *Exp. Hematol.* **1998**, *26*, 252–264. [PubMed]

19. Zhu, D.M.; Wu, L.; Suo, M.; Gao, S.; Xie, W.; Zan, M.H.; Liu, A.; Chen, B.; Wu, W.T.; Ji, L.W.; et al. Engineered red blood cells for capturing circulating tumor cells with high performance. *Nanoscale* **2018**, *10*, 6014–6023. [CrossRef] [PubMed]

20. Pezzi, H.M.; Niles, D.J.; Schehr, J.L.; Beebe, D.J.; Lang, J.M. Integration of Magnetic Bead-Based Cell Selection into Complex Isolations. *ACS Omega* **2018**, *3*, 3908–3917. [CrossRef] [PubMed]

21. Thiery, J.P. Epithelial–mesenchymal transitions in development and pathologies. *Curr. Opin. Cell Biol.* **2003**, *15*, 740–746. [CrossRef] [PubMed]

22. Mikolajczyk, S.D.; Millar, L.S.; Tsinberg, P.; Coutts, S.M.; Zomorrodi, M.; Pham, T.; Bischoff, F.Z.; Pircher, T.J. Detection of EpCAM-Negative and Cytokeratin-Negative Circulating Tumor Cells in Peripheral Blood. *J. Oncol.* **2011**, *2011*, 252361. [CrossRef] [PubMed]

23. Danila, D.C.; Heller, G.; Gignac, G.A.; Gonzalez-Espinoza, R.; Anand, A.; Tanaka, E.; Lilja, H.; Schwartz, L.; Larson, S.; Fleisher, M.; et al. Circulating tumor cell number and prognosis in progressive castration-resistant prostate cancer. *Clin. Cancer Res.* **2007**, *13*, 7053–7058. [CrossRef] [PubMed]

24. Lustberg, M.B.; Balasubramanian, P.; Miller, B.; Garcia-Villa, A.; Deighan, C.; Wu, Y.; Carothers, S.; Berger, M.; Ramaswamy, B.; Macrae, E.R.; et al. Heterogeneous atypical cell populations are present in blood of metastatic breast cancer patients. *Breast Cancer Res.* **2014**, *16*, R23. [CrossRef] [PubMed]

25. Antfolk, M.; Laurell, T. Continuous flow microfluidic separation and processing of rare cells and bioparticles found in blood—A review. *Anal. Chim. Acta* **2017**, *965*, 9–35. [CrossRef] [PubMed]

26. Hyun, K.A.; Lee, T.Y.; Lee, S.H.; Jung, H.I. Two-stage microfluidic chip for selective isolation of circulating tumor cells (CTCs). *Biosens. Bioelectron.* **2015**, *67*, 86–92. [CrossRef] [PubMed]

27. Lin, M.; Chen, J.F.; Lu, Y.T.; Zhang, Y.; Song, J.; Hou, S.; Ke, Z.; Tseng, H.R. Nanostructure embedded microchips for detection, isolation, and characterization of circulating tumor cells. *Acc. Chem. Res.* **2014**, *47*, 2941–2950. [CrossRef] [PubMed]

28. Chang, C.L.; Huang, W.; Jalal, S.I.; Chan, B.D.; Mahmood, A.; Shahda, S.; O'Neil, B.H.; Matei, D.E.; Savran, C.A. Circulating tumor cell detection using a parallel flow micro-aperture chip system. *Lab Chip* **2015**, *15*, 1677–1688. [CrossRef] [PubMed]

29. Stott, S.L.; Hsu, C.H.; Tsukrov, D.I.; Yu, M.; Miyamoto, D.T.; Waltman, B.A.; Rothenberg, S.M.; Shah, A.M.; Smas, M.E.; Korir, G.K.; et al. Isolation of circulating tumor cells using a microvortex-generating herringbone-chip. *Proc. Natl. Acad. Sci. USA* **2010**, *107*, 18392–18397. [CrossRef] [PubMed]

30. Huang, S.B.; Wu, M.H.; Lin, Y.H.; Hsieh, C.H.; Yang, C.L.; Lin, H.C.; Tseng, C.P.; Lee, G.B. High-purity and label-free isolation of circulating tumor cells (CTCs) in a microfluidic platform by using optically-induced-dielectrophoretic (ODEP) force. *Lab Chip* **2013**, *13*, 1371–1383. [CrossRef] [PubMed]

31. Chiou, P.Y.; Ohta, A.T.; Wu, M.C. Massively parallel manipulation of single cells and microparticles using optical images. *Nature* **2005**, *436*, 370–372. [CrossRef] [PubMed]

32. Chou, W.-P.; Wang, H.-M.; Chang, J.-H.; Chiu, T.-K.; Hsieh, C.-H.; Liao, C.-J.; Wu, M.-H. The utilization of optically-induced-dielectrophoresis (ODEP)-based virtual cell filters in a microfluidic system for continuous isolation and purification of circulating tumour cells (CTCs) based on their size characteristics. *Sens. Actuators B Chem.* **2017**, *241*, 245–254. [CrossRef]

33. Svoboda, K.; Block, S.M. Biological applications of optical forces. *Annu. Rev. Biophys. Biomol. Struct.* **1994**, *23*, 247–285. [CrossRef] [PubMed]

34. Ohta, A.T.; Chiou, P.Y.; Phan, H.L.; Sherwood, S.W.; Yang, J.M.; Lau, A.N.K.; Hsu, H.Y.; Jamshidi, A.; Wu, M.C. Optically controlled cell discrimination and trapping using optoelectronic tweezers. *IEEE J. Sel. Top. Quant.* **2007**, *13*, 235–243. [CrossRef]

35. Ohta, A.T.; Chiou, P.Y.; Han, T.H.; Liao, J.C.; Bhardwaj, U.; McCabe, E.R.B.; Yu, F.Q.; Sun, R.; Wu, M.C. Dynamic cell and microparticle control via optoelectronic tweezers. *J. Microelectromech. Syst.* **2007**, *16*, 491–499. [CrossRef]

36. Pethig, R. Review article-dielectrophoresis: Status of the theory, technology, and applications. *Biomicrofluidics* **2010**, *4*. [CrossRef] [PubMed]

37. Neale, S.L.; Ohta, A.T.; Hsu, H.Y.; Valley, J.K.; Jamshidi, A.; Wu, M.C. Trap profiles of projector based optoelectronic tweezers (OET) with HeLa cells. *Opt. Express* **2009**, *17*, 5232–5239. [CrossRef] [PubMed]

38. Chou, W.C.; Wu, M.H.; Chang, P.H.; Hsu, H.C.; Chang, G.J.; Huang, W.K.; Wu, C.E.; Hsieh, J.C. A Prognostic Model Based on Circulating Tumour Cells is Useful for Identifying the Poorest Survival Outcome in Patients with Metastatic Colorectal Cancer. *Int. J. Biol. Sci.* **2018**, *14*, 137–146. [CrossRef] [PubMed]

39. Lin, H.C.; Hsu, H.C.; Hsieh, C.H.; Wang, H.M.; Huang, C.Y.; Wu, M.H.; Tseng, C.P. A negative selection system PowerMag for effective leukocyte depletion and enhanced detection of EpCAM positive and negative circulating tumor cells. *Clin. Chim. Acta Int. J. Clin. Chem.* **2013**, *419*, 77–84. [CrossRef] [PubMed]

40. Chiu, T.-K.; Chao, A.C.; Chou, W.-P.; Liao, C.-J.; Wang, H.-M.; Chang, J.-H.; Chen, P.-H.; Wu, M.-H. Optically-induced-dielectrophoresis (ODEP)-based cell manipulation in a microfluidic system for high-purity isolation of integral circulating tumor cell (CTC) clusters based on their size characteristics. *Sens. Actuators B Chem.* **2018**, *258*, 1161–1173. [CrossRef]

41. Hou, S.; Zhao, H.; Zhao, L.; Shen, Q.; Wei, K.S.; Suh, D.Y.; Nakao, A.; Garcia, M.A.; Song, M.; Lee, T.; et al. Capture and stimulated release of circulating tumor cells on polymer-grafted silicon nanostructures. *Adv. Mater.* **2013**, *25*, 1547–1551. [CrossRef] [PubMed]

42. Moon, H.S.; Kwon, K.; Kim, S.I.; Han, H.; Sohn, J.; Lee, S.; Jung, H.I. Continuous separation of breast cancer cells from blood samples using multi-orifice flow fractionation (MOFF) and dielectrophoresis (DEP). *Lab Chip* **2011**, *11*, 1118–1125. [CrossRef] [PubMed]

43. Chen, L.; Zheng, X.-L.; Hu, N.; Yang, J.; Luo, H.-Y.; Jiang, F.; Liao, Y.-J. Research Progress on Microfluidic Chip of Cell Separation Based on Dielectrophoresis. *Chin. J. Anal. Chem.* **2015**, *43*, 300–309. [CrossRef]

44. Lin, Y.-H.; Lee, G.-B. An integrated cell counting and continuous cell lysis device using an optically induced electric field. *Sens. Actuators B Chem.* **2010**, *145*, 854–860. [CrossRef]

45. Huang, S.H.; Hung, L.Y.; Lee, G.B. Continuous nucleus extraction by optically-induced cell lysis on a batch-type microfluidic platform. *Lab Chip* **2016**, *16*, 1447–1456. [CrossRef] [PubMed]

46. Song-Bin, H.; Shing-Lun, L.; Jian-Ting, L.; Min-Hsien, W. Label-free Live and Dead Cell Separation Method Using a High-Efficiency Optically-Induced Dielectrophoretic (ODEP) Force-based Microfluidic Platform. *Int. J. Autom. Smart Technol.* **2014**, *4*, 83–91. [CrossRef]

47. Xie, S.; Wang, X.; Jiao, N.; Tung, S.; Liu, L. Programmable micrometer-sized motor array based on live cells. *Lab Chip* **2017**, *17*, 2046–2053. [CrossRef] [PubMed]

48. Hung, S.-H.; Lin, Y.-H.; Lee, G.-B. A microfluidic platform for manipulation and separation of oil-in-water emulsion droplets using optically induced dielectrophoresis. *J. Micromech. Microeng.* **2010**, *20*, 045026. [CrossRef]

49. Liao, C.-J.; Hsieh, C.-H.; Wang, H.-M.; Chou, W.-P.; Chiu, T.-K.; Chang, J.-H.; Chao, A.C.; Wu, M.-H. Isolation of label-free and viable circulating tumour cells (CTCs) from blood samples of cancer patients through a two-step process: Negative selection-type immunomagnetic beads and spheroid cell culture-based cell isolation. *RSC Adv.* **2017**, *7*, 29339–29349. [CrossRef]

50. Polyak, K.; Weinberg, R.A. Transitions between epithelial and mesenchymal states: Acquisition of malignant and stem cell traits. *Nat. Rev. Cancer* **2009**, *9*, 265–273. [CrossRef] [PubMed]

51. Gottesman, M.M.; Fojo, T.; Bates, S.E. Multidrug resistance in cancer: Role of ATP-dependent transporters. *Nat. Rev. Cancer* **2002**, *2*, 48–58. [CrossRef] [PubMed]

52. Shibue, T.; Weinberg, R.A. EMT, CSCs, and drug resistance: The mechanistic link and clinical implications. *Nat. Rev. Clin. Oncol.* **2017**, *14*, 611–629. [CrossRef] [PubMed]

53. Vincent, T.; Neve, E.P.A.; Johnson, J.R.; Kukalev, A.; Rojo, F.; Albanell, J.; Pietras, K.; Virtanen, I.; Philipson, L.; Leopold, P.L.; et al. A SNAIL1-SMAD3/4 transcriptional repressor complex promotes TGF-beta mediated epithelial-mesenchymal transition. *Nat. Cell Biol.* **2009**, *11*, 943–950. [CrossRef] [PubMed]

54. Satelli, A.; Li, S.L. Vimentin in cancer and its potential as a molecular target for cancer therapy. *Cell. Mol. Life Sci.* **2011**, *68*, 3033–3046. [CrossRef] [PubMed]

55. Kaufmann, O.; Deidesheimer, T.; Muehlenberg, M.; Deicke, P.; Dietel, M. Immunohistochemical differentiation of metastatic breast carcinomas from metastatic adenocarcinomas of other common primary sites. *Histopathology* **1996**, *29*, 233–240. [CrossRef] [PubMed]

56. Smith, A.; Teknos, T.N.; Pan, Q.T. Epithelial to mesenchymal transition in head and neck squamous cell carcinoma. *Oral Oncol.* **2013**, *49*, 287–292. [CrossRef] [PubMed]

57. Kallergi, G.; Papadaki, M.A.; Politaki, E.; Mavroudis, D.; Georgoulias, V.; Agelaki, S. Epithelial to mesenchymal transition markers expressed in circulating tumour cells of early and metastatic breast cancer patients. *Breast Cancer Res.* **2011**, *13*. [CrossRef] [PubMed]

58. Poruk, K.E.; Valero, V.; Saunders, T.; Blackford, A.L.; Griffin, J.F.; Poling, J.; Hruban, R.H.; Anders, R.A.; Herman, J.; Zheng, L.; et al. Circulating Tumor Cell Phenotype Predicts Recurrence and Survival in Pancreatic Adenocarcinoma. *Ann. Surg.* **2016**, *264*, 1073–1081. [CrossRef] [PubMed]

59. Satelli, A.; Mitra, A.; Brownlee, Z.; Xia, X.; Bellister, S.; Overman, M.J.; Kopetz, S.; Ellis, L.M.; Meng, Q.H.; Li, S. Epithelial-mesenchymal transitioned circulating tumor cells capture for detecting tumor progression. *Clin. Cancer Res.* **2015**, *21*, 899–906. [CrossRef] [PubMed]

60. Lindsay, C.R.; Le Moulec, S.; Billiot, F.; Loriot, Y.; Ngo-Camus, M.; Vielh, P.; Fizazi, K.; Massard, C.; Farace, F. Vimentin and Ki67 expression in circulating tumour cells derived from castrate-resistant prostate cancer. *BMC Cancer* **2016**, *16*, 168. [CrossRef] [PubMed]

61. Chen, Y.C.; Chen, Y.W.; Hsu, H.S.; Tseng, L.M.; Huang, P.I.; Lu, K.H.; Chen, D.T.; Tai, L.K.; Yung, M.C.; Chang, S.C.; et al. Aldehyde dehydrogenase 1 is a putative marker for cancer stem cells in head and neck squamous cancer. *Biochem. Biophys. Res. Commun.* **2009**, *385*, 307–313. [CrossRef] [PubMed]

62. Snyder, V.; Reed-Newman, T.C.; Arnold, L.; Thomas, S.M.; Anant, S. Cancer Stem Cell Metabolism and Potential Therapeutic Targets. *Front. Oncol.* **2018**, *8*, 203. [CrossRef] [PubMed]

MDPI

St. Alban-Anlage 66

4052 Basel

Switzerland

Tel. +41 61 683 77 34

Fax +41 61 302 89 18

www.mdpi.com

Micromachines Editorial Office

E-mail: micromachines@mdpi.com

www.mdpi.com/journal/micromachines